A Field Guide to
Airplanes of North America

A Field Guide to

AIRPLANES

of North America

THIRD EDITION

M. R. Montgomery
and Gerald L. Foster

Illustrated by Gerald L. Foster

 HOUGHTON MIFFLIN COMPANY
Boston   New York

Library of Congress Cataloging-in-Publication Data
Montgomery, M. R.
A field guide to airplanes of North America / M. R. Montgomery and Gerald L. Foster ;
illustrated by Gerald L. Foster. — 3rd ed.
p. cm.
Includes index.
ISBN-13: 978-0-618-41127-6
ISBN-10: 0-618-41127-5
1. Airplanes — Recognition. 2. Helicopters — Recognition. I. Foster, Gerald L. II. Title.
TL671.M64 2006
629.133'34 — dc22   2005010335

CONTENTS

INTRODUCTION

The purpose of this field guide has always been simple: to allow people of any age to *identify* any factory-built aircraft and all the military aircraft they are likely to see anywhere in North America, from Mexico to the Arctic Circle. For the plane watcher, the rise of the Internet has created an enormous volume of very accessible and very detailed information on particular aircraft, but *specific information is accessed efficiently only after you know the manufacturer and the model.*

This third, and greatly expanded, edition of *A Field Guide*...adds more than eighty new aircraft that have appeared since the 1992 edition. The 1980s was an era of stagnation in general aviation. The industry had been decimated by increasing liability insurance costs; one venerable company, Piper Aircraft, went out of business, and another, Cessna, got out of the private market to concentrate on more profitable business aircraft, especially jets. Several of the old Piper designs (with expired liability) are produced by the New Piper Corporation, based in Florida.

Two changes have allowed the old companies to revive: Reform in liability litigation has been helpful, and the case-law decision that any aircraft that had been certified more than twenty-five years was by its nature not a design that would be, *as a design,* a source of liability. In that long down period for U.S. manufacturers, a number of foreign-built general-aviation airplanes established themselves in the United States, in some cases by building aircraft aimed precisely at gaps created when the major manufacturers got out of the business of selling to the private citizen.

Further, an explosion of short- to medium-range airliners, all imported, has filled the huge gaps left when U.S. manufacturers stretched their small airliners from regional jets carrying several dozen passengers to planes seating in excess of one hundred. At this point, the commonest aircraft at many jetports are Canadian and Brazilian in origin, and they are treated in detail in this new edition. New helicopters abound, particularly for serving off-shore oil platforms and doing both rescue and recreation flights into very high and difficult terrain.

Although not the only guide to aircraft identification in the history of publishing, this one is unique. Like the Peterson Field Guides, this guide is devoted to a single geographical area. Although many of the aircraft in this book are seen throughout the world, we have excluded foreign aircraft that are never, or very rarely, seen in North American airspace. Guides to "all the world's" aircraft eliminate all the older and rarer planes, lump complex series into an indistinct and blurred composite aircraft, and confuse readers by showing them Russian or European planes that simply do not operate in our skies. Just as bird guides do not include all the fauna in zoological parks, this guide does not include every museum piece. This book is about what's up there now, and we know it is also useful worldwide, particularly for older aircraft during the era when the United States dominated the industry.

With very few exceptions, at least thirty examples of a type must be still flying for it to be included in the field guide. In the case of civil aircraft, if you can buy a ticket to fly, as either an airline passenger or a tourist on a scenic flight, we have included the aircraft. Flying concentrates the mind and provokes interest in the hardware. Military aircraft are so bizarre (and noisy) that we include all currently operational military aircraft, including the most obscure survivors, even if only a handful are flying. And we confess to including some planes that you will be lucky to find but that are so charming or so important in the development of subsequent aircraft that we decided to include our favorites. These few are restricted to propeller-driven aircraft and aren't so many as to create a problem for the quick and simple identification of the more numerous planes.

One single class of fixed-wing aircraft is not fully covered: the home-builts. Their variety is too great, and builders may modify them to suit their own tastes. However, several of the planes included here have been both factory- and home-built, and that is noted in the text. In particular, we have covered the most popular home-built biplanes because they are patterned after production aircraft of the 1930s and 1940s.

How to Use This Book

Aircraft are grouped by both their use and their appearance. The book begins with the small airplanes in what is usually called *general aviation*, starting with biplanes and followed by agricultural planes (including agricultural biplanes). Single-engine propeller-driven planes are grouped by such quickly visible field marks as whether they have wings mounted on top of the fuselage or at the bottom; by landing gear, fixed or retractable; by type of gear, tail dragging or tricycle. Several manufacturers have made essentially the same plane with fixed or retractable gear, and these planes are grouped in the transitional pages between types of aircraft.

Both the multiengine props and jets are grouped by size. Despite a certain charm to keeping all twin, fuselage-mounted, swept-wing jets together, that would have put aircraft as large as a stretched MD80 carrying nearly two hundred passengers next to the much smaller, not really similar, Falcon 20 business jet that seats eight.

Special-purpose military aircraft—combat, transportation, observation—are grouped together. However, dozens of commercial and general-aviation aircraft—planes and helicopters—are acquired by the military for transportation, often of VIPs. Pure military planes and helicopters simply don't look like jetliners or business jets. Something conventional-looking but wearing military insignia or camouflage can be easily identified by looking for it in the appropriate civilian section of the book.

We have avoided technical language whenever possible and would just as soon think of *vertical stabilizers* as tail fins and call them that. However, some useful field marks have their own aeronautical terms: We should define *chord, dihedral, fairing,* and *nacelle*.

The best way to describe a wing that is the same width along its entire length is to refer to its *constant chord* (from the geometry term describing the distance across the bottom of a curve, measured in a straight line; all wings are curved across the top).

Another useful technical term is *dihedral,* which describes wings or tail planes (*horizontal stabilizers*) that angle upward, so that the wing tip is elevated above the root of the wing at the fuselage. Even very slight dihedrals in small tail planes, as well as in long wings, are quite noticeable from a distance.

The word *fairing* appears often and is an old word from ship's architecture adapted to aeronautics. A fairing is simply a smoothed-out or streamlined connection between two parts of a vessel—ship or aircraft—that often conceals structural bracing. Fairings are common at wing roots and where engines are inserted into wings. The engine housings are called *nacelles* (from an old French word meaning *little boat,* which captures the tapering shape rather nicely).

Identifying a particular aircraft usually requires recognizing a combination of two or more field marks. For some similar models, you may be reduced to counting passenger windows or noting the shape of the windows. The easiest place to identify planes is at an airport, just as the easiest place to identify birds is at a near-at-hand bird feeder. And, as happens with birding, once you have made a positive identification of a perching bird and then have seen it fly, some of the little field marks become irrelevant, and you recognize the bird as a whole, not just as the sum of its field-mark parts. British birders, for reasons unclear, refer to this as the *jizz* of the bird. We like to think of it as the (hard *g*) *gestalt,* the German word for the unique presence of a person or a thing. A stretched DC8, once you have seen one nearby and then watched it disappear into the distance, will always be instantly identifiable at any range a long, skinny fuselage balanced on relatively small wings.

There is no rigid order for using the field marks. We suggest that you thumb through the sections of the book; get a sense of where the high-wings and low-wings, propellers and jets, fixed and retractable airplanes are located; and browse the field marks for a variety of aircraft *before* you start to use the book. Get a sense of the useful field marks, and try to find them all at once—this will work much better than some rigid litany of "wing, tail, landing gear, window..." As with any field guide, familiarity with the book is the best system.

A Field Guide to
Airplanes of North America

Beech 17 Staggerwing (Navy GB-1, Air Force C-43)
Length: 26'9" (8.13 m) **Wingspan:** 32' (9.76 m) **Cruising speed:** 201 mph (323 km/h)

Rare. Large; reversed staggerwing (lower wing forward of upper); enclosed cabin; solid wing struts.

The Rolls-Royce of biplanes. Performance data is for the most powerful versions, with 450 hp engines. First flown in 1932 with fixed landing gear; never seen today without the electrically operated retractable gear. Various models have slight dimensional changes, but all are clearly Staggerwings. Once a popular float and ski plane. A few postwar models, last produced in 1948, have leather upholstery and other comforts.

Note: Any cabin biplane that is not a Beech 17 (reversed staggerwing) is a Waco.

Any cabin biplane with an upper wing much longer and deeper than the bottom wing is a late-model Waco C (custom) biplane.

All other cabin biplanes with wings of equal width and normal stagger are Waco S (standard) or very early C (custom) planes.

Waco Late C Series
Length: 27'7" (8.42 m) **Wingspan:** upper, 34'9" (10.57 m); lower, 24'6" (7.47 m)
Cruising speed: 155 mph (249 km/h)

Rare. Cabin biplane with a noticeably shorter and narrower lower wing (compare with Waco S series, next entry); fixed landing gear; N wing struts, plus a heavy brace from the base of the N strut to the upper wing.

One of four basic types of Waco biplanes, the late C (custom cabin) series is the only one with the very small, normally staggered lower wing. Built throughout the 1930s. The fixed gear is usually seen with streamlined wheel pants. Proper restoration includes the straight-line striping from the engine cowling to the tail plane. A few Waco biplanes were in U.S. and foreign military service, but for the famous WWII basic trainer, see the Waco UPF7 (page 4).

Waco S Series, Early C Series
Length: 25'3" (7.71 m) **Wingspan:** upper, 33'3" (10.15 m); lower, 28'3" (8.62 m)
Cruising speed: 133 mph (214 km/h)

Rare. Cabin biplane with slightly shorter lower wing; wings of equal width (chord); N struts, plus solid brace.

The S (standard) and early C (custom) Waco biplanes are handsome, symmetrical, and remarkable for their lack of unusual features. They have very similar upper and lower wings, typical struts, and a conventional cabin. Usually restored with the Waco signature stripe from cowling to tail. Both wings have a matching, very slight dihedral. Although they were not supplied with streamlined wheel pants, you may see one that's been modified. Concentrate on the wings.

Beech 17 Staggerwing

Waco Late C Series

Waco S Series

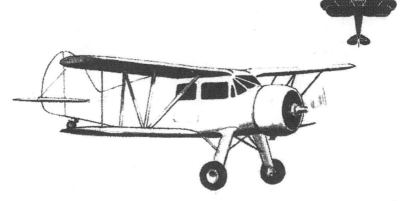

Boeing/Stearman Kaydet (military PT-13, PT-17, PT-18)
Length: 24'10" (7.58 m) **Wingspan:** upper, 32'2" (9.82 m); lower, 1' shorter overall
Cruising speed: 103 mph (166 km/h)

Fairly common. The normally staggered wings of almost equal length, combined with the unbraced heavy landing gear and the N struts without an aileron connector, separate the Kaydet from the somewhat similar biplanes of the 1930s and 1940s. Compare with the aircraft in the next three entries.

More than 10,000 Stearmans were built from the early 1930s through WWII; model designators indicate engines of different horsepower. A jointly procured trainer for the Navy and the Army Air Corps, many are seen restored to their WWII paint scheme: Air Force blue fuselage and Navy yellow wings with service markings. Note that although the cockpits are large and deep, there is no turtleback behind the rear cockpit.

Naval Aircraft Factory N3N1, N3N3
Length: 25'11" (7.96 m) **Wingspan:** 34' (10.38 m) **Cruising speed:** 92 mph (148 km/h)

Rare. Normally staggered wings identical in length and width (chord); N struts with aileron connector; skinny braced landing gear without wheel pants; no engine cowling.

Once used extensively as agricultural aircraft, the government-built N3Ns are collectors' items. A proper restoration is all yellow with Navy insignia. The last biplane in U.S. service, until 1958, as a float plane at the U.S. Naval Academy, Annapolis. All midshipmen, whether they were aviators or not, had to spend 10 hrs. flying in the "Yellow Peril," an experience that, for many, was equaled for sheer terror only by submarine-escape training.

Waco UPF7, YPF7 (military trainer PT-14), Model D
Length: 23'1" (7.06 m) **Wingspan:** 30' (9.14 m); lower, 26'10" (8.18 m)
Cruising speed: 123 mph (198 km/h)

Fairly common. Lower wing noticeably shorter; look for the large rectangular cutout in the upper wing; designed for easier access to the forward cockpit; longer nosed than the early F series; may or may not have engine cowling.

Although a military trainer in WWII, not as common as the Stearman Kaydets or the Naval Factory N3N series. Very popular primary trainer with the WWII government Civilian Pilot Training Program. A sports type (Waco model D) was built with streamlined wheel pants and lighter construction materials.

Boeing/Stearman Kaydet

Naval Aircraft Factory N3N3

Waco UPF7, PT-14

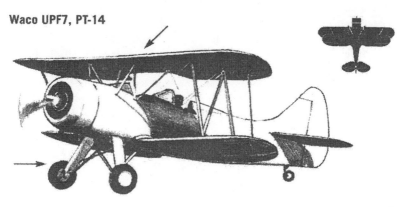

Classic Aircraft Waco F-5

Length: 23'4" (7.10 m) **Wingspan:** upper, 30' (9.14 m); lower, 26'10" (8.18 m)
Cruising speed: 110 mph (198 km/h)

Least common but newest of the F series. **Lower wing slightly shorter than upper; engine cowling shows bumps, never smooth; always with streamlined wheel pants;** up close, can be distinguished from the old F series by the **added navigation lights.**

Built under the original Model F-5 certification but with modern corrosion-proof metals, hydraulic brakes, self-starting engine, and fireproofing forward of the engine wall. Classic Aircraft, of Lansing, Michigan, produced more than 40, now scattered on airfields from Hawaii to Maine. Rarest in the Pacific Northwest (where open cockpits are wet cockpits).

Waco Early F Series

Length: 20'9" (6.31 m) **Wingspan:** 29'6" (9 m) **Cruising speed:** 90 mph (145 km/h)

Rare. May be confused with the Waco UPF7 or the naval aircraft trainer **but very stubby nosed; wings of equal length; N brace with aileron connector; small circular cutout in top wing** for access to front cockpit; distinct turtleback behind rear cockpit.

A popular sportster and trainer from early 1930s, the early F series is popular with restorers but much less common than the Waco UPF7 military trainers, which it slightly resembles. Built with and without engine cowlings—some with ring cowlings and some with streamlined cowlings—but typically with exposed radial engine cylinder heads. Landing gear usually bare.

Travel Air 4000

Length: 24'2" (7.35 m) **Wingspan:** 34'8" (10.53 m) **Cruising speed:** 100 mph (161 km/h)

Rare. **Looks distinctly antique;** almost always shows the **elephant ear upper wingtip and tail fin; N bracing,** plus aileron control transfer bar; some built with conventional speed wings, but these show the elephant ear tail; a few with conventionally rounded tails, but these always show the upper-wing elephant ear, which is an extension of the aileron; both wings straight, with lower wing noticeably shorter and slightly narrower.

The Travel Air was built in a variety of versions, including passenger carriers, with a two-person forward cockpit. All originals and accurate restorations have either radial (in the more numerous 4000 series) or in-line (in the very rare 2000 series) water-cooled engines. A small radiator extends below the fuselage, just forward of the cockpit area. The high, quickly rising turtleback is unique.

Classic Aircraft Waco F-5

Waco Early F Series QCF2

Travel Air 4000

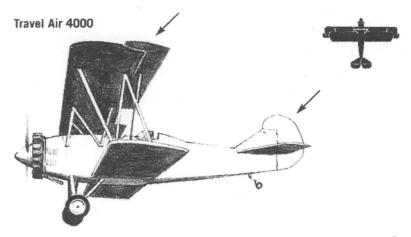

Fleet Finch Trainer

Length: 21'8" (7.1 m) **Wingspan:** both, 28' (8.53 m) **Cruising speed:** 98 mph (158 km/h)

Rare. Very stubby nosed; straight wings of equal length; lower wing with noticeable dihedral; N bracing; no aileron control transfer bar. Early models, built in the United States, have elephant ear tails.

Made in the United States in the early 1930s, then in Canada, where more than 600 were built from 1938 to 1941 for Royal Canadian Air Force flight training. Many restored Canadian-built WWII trainers have a single sliding canopy that covers both cockpits; other models have simple, flat-glass, three-sided windshields. Once a popular ski and float plane.

Meyers OTW (Out to Win)

Length: 22'8" (6.91 m) **Wingspan:** both, 30' (9.14 m) **Cruising speed:** 100 mph (161 km/h)

Rare. Combines **all-aluminum fuselage** with **fabric wings;** wings are identical, with slight dihedral; **the landing gear strut shock-absorbing piston,** which extends up to the forward cockpit, **is diagnostic.**

During WWII, only 102 OTWs were built, all for the Civilian Pilot Training Program, and half of them are still registered: some flying; the others being restored. Their use as crop-dusters after WWII contributed to the loss of many of the aircraft. Manufactured in Romulus, Michigan, from 1940 to 1944 by people who had never before, and never afterward, built airplanes.

de Havilland DH82 Tiger Moth, PT-24

Length: 23'11" (7.29 m) **Wingspan:** 29'4" (8.94 m) **Cruising speed:** 90 mph (145 km/h)

Fairly common for an antique biplane. **Swept wings of equal length; stout double-bar wing connectors (not N);** the entire plane gives a distinct impression of slimness, including the in-line engine and the fancifully tapered tail fin and tail planes.

The Tiger Moth, a 130 hp version of the 1920s Gipsy Moth, first flew in 1932 and was produced through WWII, totaling more than 8,000 planes. The standard Royal Air Force and Royal Navy primary trainer; a few hundred in U.S. Army Air Forces, designated PT-24. Surplus Moths were the backbone of private aviation in Great Britain and Canada after WWII. The bulge where the upper wing crosses the fuselage over the cockpit is the fuel tank.

Fleet Finch Trainer

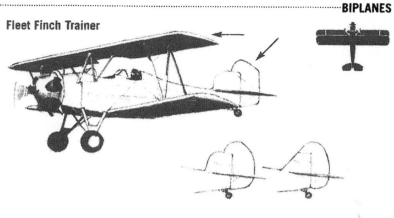

Meyers OTW

de Havilland DH82 Tiger Moth

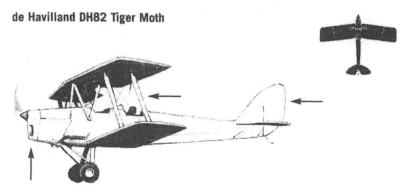

Great Lakes Sport Trainer, Baby Lakes
Great Lakes data: Length: 20'4" (6.2 m) **Wingspan:** 26'8" (8.13 m)
Cruising speed: 110 mph (177 km/h). **Baby Lakes data: Length:** 13'9" (4.10 m)
Wingspan: 16'8" (5.08 m) **Cruising speeds:** various, depending on optional engines

The original Great Lakes, built between 1929 and 1932, and the revival, built between 1974 and 1978, were tandem dual controls; the Baby Lakes is ⁵⁄₁₀ their size and is either single or dual. They share the identifying combination: **top wing swept, over straight bottom wing and N struts.** Owners can modify struts to a single one, thereby possibly causing confusion with the Pitts Special (next entry). Call it a Pitts/Lake, especially if the wheel struts on the Great Lakes have been covered with streamlining sheet metal. Original Great Lakes had ailerons on the lower wing only; some planes have been modified and show the aileron transfer control bar next to the N brace.

Although only 200 of the original Great Lakes trainers were built, they dominated acrobatics and closed-course racing in the United States in the 1930s. The company was revived and several Great Lakes versions, of greatly varying horsepower, were built. You may even see a one-seat, full-size Great Lakes. Concentrate on the wing and wing strut combination. It's unique.

Aviat Pitts S-1, S-2 Special
S-1 data: Length: 15'5" (4.7 m) **Wingspan:** 17'4" (5.28 m) **Cruising speed:** 140 mph (225 km/h)

Usually seen in the S-1 (single-seat) version, a chunky little plane. The unique combination is **top wing swept and slightly longer than straight lower wing; single wing strut plus aileron control transfer bar. Optional fuselage/upper-wing bracing may originate from two points on the wing rather than the typical N bracing.** The turtleback is high and distinctive.

The single-seat S-1 is unique in that it is available as a factory-built and certified plane or as plans or kits. The S-2 dual-control model is available only through the factory. These planes have been flown with all manner of engines, up to 450 hp, and became the premier aerobatic airplane in the 1960s. Home-built Pitts Specials may show additional bracing and wiring, probably out of a deep sense of insecurity on the part of the builder.

Aviat Christen Eagle I, II
Eagle II (two-seater) data: Length: 18'6" (5.64 m) **Wingspan:** 19'11" (6.07 m)
Cruising speed: 158 mph (254 km/h)

A kit builder's plane. The one-seat Eagle I, introduced in late 1982, has **both wings swept, single strut, and bubble canopy.** It's almost always seen with Eagle paint job; **long-nosed, large propeller spinner.**

Great Lakes Sport Trainer

Baby Lakes

Aviat Pitts S-2 Special

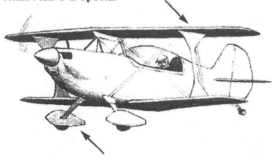

Aviat Christen Eagle II

Stolp Starduster, Acroduster

Starduster 100 data: Length: 16'6" (5.03 m) **Wingspan:** upper, 19' (5.79 m); lower, 18' (5.49 m) **Cruising speed:** 132 mph (212 km/h)

A family of home-builts. The Stardusters and the more strongly constructed aerobatic Acrodusters have **unequal-span wings.** Only the **upper wing is swept;** single interplane strut and aileron transfer control bar, **fully rounded wingtips.** Also seen in two-seaters; separates from same-sized Christen Eagles by the asymmetry of the wings. See the similar Steen Skybolt (next entry) and note its less rounded wingtips.

Steen Skybolt

Length: 19' (5.79 m) **Wingspan:** upper, 24' (7.32 m); lower, 23' (7.01 m) **Cruising speed:** 130 mph (209 km/h)

Always a two-seater. **Upper wing swept; lower, straight. Very long nosed; large, rounded tail fin. Wing braces over the fuselage radiate from two points on the wing.** Compare with the more conventional combination N braces on a Stolp Starduster. Sold as plans, with wing and fuselage kits available. More than 2,500 kits have been sold.

Smith Miniplane

Length: 15'3" (4.65 m) **Wingspan:** upper, 17' (5.18 m); lower, 15'9" (4.80 m) **Cruising speed:** 118 mph (190 km/h)

Properly called "mini." Small size; **wings not swept; lower wing slightly shorter; conventional N bracing.** The first models were known as DSA-1 (for Darn Small Airplane). Compare with the very similar EAA Biplane (next entry). EAAs tend to have a more streamlined engine cowling and a more upright tail fin.

EAA Biplane

Length: 17' (5.18 m) **Wingspan:** both, 20' (6.10 m) **Cruising speed:** 110 mph (177 km/h)

A small single-seater with **unswept, equal-length wings** and conventional **N struts.** A subtle difference between the EAA Biplane and the Smith Miniplane is the way the lower wing appears to come out of the EAA fuselage; in the Smith Mini, the fuselage appears to sit on top of the wing. The Smith Mini has a noticeably shorter lower wing.

EAA Acro-Sport, Acro-Sport II

Acro-Sport (single-seater) data: Length: 17'6" (5.33 m) **Wingspan:** upper, 19'7" (5.97 m); lower, 19'1" (5.82 m) **Cruising speed:** 105 mph (169 km/h)

The only biplane illustrated here with **unswept wings of nearly equal length and a single streamlined strut, plus aileron control transfer bar.** Designed to be built from plans and construction manuals. More than 800 have been built and flown.

Stolp Starduster

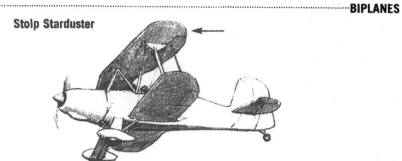

Steen Skybolt

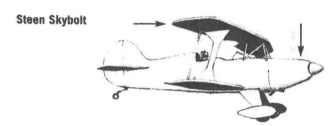

Smith Miniplane

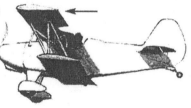

EAA Biplane

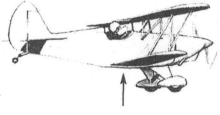

EAA Acro-Sport

✈ AGRICULTURAL PLANES

Eagle Aircraft Eagle 220, 300

Length: 27'6" (8.38 m) **Wingspan:** 55' (16.76 m) **Working speed:** 65–115 mph (105–185 km/h)

Not common. Introduced in 1981, with more than 90 produced by mid-1983. A biplane with **extremely long, thin wings.** The typical agplane cockpit sits amid a maze of wires, struts, and braces; **large tail fin.**

A revival from the era when biplanes dominated the agricultural spraying industry, this Bellanca-designed agplane has an aspect totally different from the old biplanes converted to spraying. The wings are based on sailplane designs: long, thin, and tapering. Earliest versions (not illustrated) used a radial engine, and the total length was only 26 ft. (7.92 m). Last models produced were in-line pistons; model numbers (220, 300) indicate horsepower.

Schweizer (Grumman) Ag-Cat

Length: 25'7" (7.80 m) **Wingspan:** 42'3" (12.88 m) **Working speed:** 98 mph (158 km/h)

Distinguish this biplane/agplane from older biplanes converted to crop use by its **massive, high tail fin; all-metal skin; modern roll-bar cockpit; and trimmed speed-wing wingtips.**

The original Ag-Cat was designed by Grumman but never manufactured until Schweizer, a family-run designer of sailplanes, started manufacturing it under license from Grumman in 1957. Since 1981, Schweizer has been the sole owner of the design, now marketing an Ag-Cat B with a standard 600 hp radial engine. When one considers this class of agricultural planes—many (like the Schweizer) with pressurized cockpits to keep aerial sprays and dusts away from the pilot, air conditioning, and airframes meant to collapse slowly around a rigid cockpit in the case of a crash—one ceases to wonder why there are very few old, bold crop-dusters. Compare these planes with the Call-Air A2 (page 16), whose pilot simply put a barrel of pesticide in the passenger's seat and took off.

Schweizer Ag-Cat Super-B

Length: 24'5" (7.44 m) **Wingspan:** 42'5" (12.93 m) **Working speed:** 115 mph (185 km/h)

Very similar to the Schweizer (Grumman) Ag-Cat when fitted with radial engine, but **the increased distance between the upper and lower wings is quite noticeable.** Even when fitted with a turbine engine (bottom sketch), this plane can't be confused with the Eagle (top of page), because of its shorter, broader wings and massive tail.

Although the radial Super-Bs look like their Schweizer (Grumman) ancestors, of which some 2,000 fly worldwide, the small but visible change in the top wing, raising it 8 in. (20 cm), improved the plane's lift by decreasing wing-to-wing turbulence and enhanced the pilot's forward vision while diving and upward vision in all attitudes.

Eagle Aircraft Eagle 300

Schweizer Ag-Cat

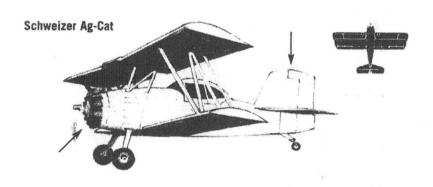

Schweizer Ag-Cat Super-B

Call-Air A2, A5
Length: 23'5" (7.25 m) **Wingspan:** 36' (11.11 m) **Cruising speed:** 102 mph (164 km/h)

Extremely rare and probably permanently parked in a quiet part of the airfield. The only production passenger aircraft with a low, braced wing. Wing is constant chord (width) with rounded tips; three-strut landing gear usually has two struts covered with speed pants. Compare with the Intermountain Call-Air A9 agricultural plane (next entry).

Fewer than 50 built as passenger planes, a few more as Call-Air A5 and A6 crop-dusters, with spray material carried inside the A2-style cabin; included here because its use of the constant-chord wing with high-lift qualities was unique when the plane was designed in 1939. Built in Wyoming at an airfield with an elevation of 6,200 ft., the Call-Air was perfectly at home in "high and hot" thin air.

Intermountain Mfg. Co. Call-Air A9, Aero Commander, Sparrow, Quail, Snipe, AAM Thrush Commander
Length: 24' (7.32 m) **Wingspan:** 35' (10.67 m) **Working speed:** 100 mph (161 km/h)

Not so common as some agricultural planes; last produced in Mexico by Aeronautica Agricola Mexicana. Typical agplane shape, low wing braced with three struts, equal-chord (width) wings, light wire braces on tail planes, triple braces to forward wheels, somewhat old-fashioned curved tail fin and tail planes. A rare Snipe model has a radial engine.

Agplane fans will see the family history of the Call-Air A9 in the triple wing braces and triple wheel struts, picked up from the original Call-Air A2 monoplane (previous entry) and the now very rare Call-Air A5 and A6 agplanes. A Wyoming company developed the Call-Air A9 and manufactured a few hundred from 1963 to 1965. That design was sold to Aero Commander, a division of Rockwell—later, North American Rockwell. The A9 design survives today in the triple braces to the front wheels in the Thrush agplanes, which have a modern, unbraced wing. Rockwell sold the braced-wing design to Aeronautica Agricola Mexicana. A few of the earliest Call-Air A9s did not have windows in the roof of the cockpit. Close at hand, note the distinct droop to the leading edge of the wing, giving the plane a very short takeoff roll (1,200 ft.) when fully loaded.

Call-Air A2, A5

Intermountain Call-Air A9

Aero Commander Quail

Piper PA25 Pawnee

Length: 24' (7.32 m) **Wingspan:** 36'2" (11.02 m) **Working speed:** 95 mph (153 km/h)

A small, old-fashioned-looking agplane is either a Pawnee or one of the Sparrow Commander/Call-Air A9 types; compare with them before deciding. **Low wing has a pair of braces on top; tail planes have paired braces top and bottom; wings are fabric over rib; and it usually shows up clearly rounded wingtips and rounded tail geometry.**

One of the first pure agplanes; built between 1959 and 1982; early replacement for the old biplane dusters. The high placement of the pilot, the rear cockpit windows, and the extra-long nose for progressive collapse if crashed, as well as interior safety features, were designed with the assistance of Cornell University agricultural and mechanical engineering studies.

Cessna Ag Truck, Ag Wagon, Ag Pickup, Ag Husky

Length: 25'3" (7.70 m) **Wingspan:** 40'4" (12.30 m) **Working speed:** variable, about 100 mph (161 km/h)

Quite variable window configurations but always with these constants: **Wing is braced by a single, streamlined strut that is faired into the wing; unbraced tail planes; single, spring-steel struts to front wheels; very sharp (9-degree) dihedral that begins after the wing leaves the fuselage horizontally.**

Developed in 1965, the Cessna Ag series has a number of names signifying nothing more than varieties of engines, load-carrying capacity, and variations in windows; many early models before 1969 lacked the rear and top cockpit windows. Beginning in 1971, a few models had high-lift drooped wingtips. Since 1980, all models (and other Cessna singles) have the conical camber wingtips.

Piper PA36 Brave, Pawnee Brave, WTA New Brave 375, 400

Length: 27'6" (8.38 m) **Wingspan:** 38'9" (11.83 m) **Working speed:** 112 mph (180 km/h)

Typical agricultural low-wing monoplane. **Unbraced wings** (compare with Thrush and Air Tractor, next two entries); **wings of equal chord** (after fairing at wing root); **unbraced tail plane; streamlined forward landing gear struts; shock absorbing; squared-off shape to tail fin, wingtips, and tail planes.**

Developed by Piper in 1972; once manufactured by WTA, Inc., a Texas company that also produced a Piper PA18 Super Cub. The extralong nose of the Brave is so designed to collapse progressively in case of a crash. Not manufactured with radial engines or in two-seat models.

Piper PA25 Pawnee

Cessna Ag Truck

Piper PA36 Brave

Ayres Thrush, Bull Thrush, Turbo Thrush, Rockwell-Commander Thrush
Length: 29'5" (8.96 m) **Wingspan:** 44'5" (13.54 m) **Working speed:** 110 mph (177 km/h)

Typical agricultural low-wing monoplane with **unbraced wings;** compare with the Air Tractor (next entry) before deciding; **fixed gear; three struts for each forward wheel; pair of thin wire braces above and below tail planes; equal-chord (width) wings, with trapezoidal tips.**

Developed by Rockwell-Commander in 1965; manufactured by the Ayres Corporation after 1977. Comes in various configurations, but all have the same field marks. The original models came with radial engines; recently, with in-line turboprop engines (upper sketch). A two-seat cabin is standard on the 1,200 hp radial Bull Thrush (lower sketch) but is also available on the turboprop airframe. Bull Thrush carries up to 510 gal. liquid spray.

Air Tractor 301, 401, 501, 400, 402, 502, 503
Length: 27' (8.23 m) **Wingspan:** 45'1" (13.75 m) **Working speed:** 130 mph (209 km/h)

Typical low-wing agricultural plane. Unbraced wing (compare with the Thrush [previous entry] before deciding); **fixed gear; single spring-steel strut carries each wheel; wing of equal chord (depth), with straight squared-off wingtips; pair of light braces on the underside only of the tail plane.**

Manufactured in various models since 1972. The field marks are consistent, although the plane is equipped with radial engines (model 301, lower sketch) or turboprop engines (models 302, 400, 402, 502, upper sketch). Designed by Leland Snow, who also designed the Snow S2 agplanes, which became the Rockwell Thrush, now the Ayres Thrush. It is also manufactured in a two-seater (compare with the Thrush).

Weatherly 620, 620TP, 201
Length: 27'3" (8.30 m) **Wingspan:** 41' (12.5 m) **Working speed:** 105 mph (169 km/h)

Not common and quite variable. All models have **low, unbraced wing of constant chord (width); very strong dihedral begins a few feet out from fuselage; top of triangular tail fin is clipped.** An option is detachable vanes that extend the spray path by about 8 ft. (2.47 m).

Weatherly Aviation began by converting Fairchild M62s (page 22) to crop sprayers and continued with its own modifications of that design. Except for the radial engines on some models (lower sketch), the plane's air of angularity — including the constant-chord wings, the delta tail fin, and the trapezoidal tail planes — is unique. Even the tapers in the fuselage section appear to be flat sections.

Ayres Turbo Thrush

Ayres Bull Thrush

Air Tractor 400, 402, 502

Air Tractor 301, 401, 501

Weatherly 620TP

Weatherly 620, 201

21

LOW-WING SINGLES

Ryan ST3 (PT-21, PT-22 NR-1), Ryan ST
Length: 22'5" (6.83 m) **Wingspan:** 30'1" (9.18 m) **Cruising speed:** 123 mph (198 km/h)

Quite rare. Constant-chord low wing; rounded tips; both the wings and the tail planes are braced, top and bottom, with wire; **cylinder heads of the standard engine project through cowling; distinct, abrupt turtle-back to rear cockpit.**

Of the thousands built, more than 500 PT-21s survived WWII training duties and entered the civilian market. Although slow, the plane was more than strong enough for acrobatics (the point of the noisy wire bracing). The plane had a fairly high stall speed, 64 mph (103 km/h), and sank like a rock without power. The civil version (ST) had an in-line engine and wheel pants (see sketch); the military five-banger was easier to work on, and the wheel pants were dropped in deference to the abuse that landing gears took from student pilots.

Fairchild PT-19 (M62), Cornell
Length: 27'8" (8.5 m) **Wingspan:** 35'11" (11 m) **Cruising speed:** 120 mph (193 km/h)

Rare old birds. **Unbraced low wing; twin tandem cockpits (which may be enclosed in a greenhouse, upper sketch); fixed tail-dragger landing gear without wheel pants.**

Built by the thousands; a largely wood spar and plywood-exterior basic trainer flown by nearly a million WWII student pilots. Faster and sturdier than the biplanes of that era. When fitted with radial engines, known as the PT-23 (lower sketch) —a much less common type than the PT-19 (middle sketch). Greenhouse canopy supplied on Canadian Air Force versions (the Cornell) and on the few civilian models, designated M62. All were remarkably durable (although the wood construction has created problems after the passage of nearly 50 years) and regarded as forgiving and easy to fly.

Consolidated Vultee Valiant, BT-13, BT-15, SNV-1
Length: 28'7" (8.65 m) **Wingspan:** 42' (12.8 m) **Cruising speed:** 170 mph (274 km/h)

Quite rare, although 10,000 built through WWII. An odd combination: **fully enclosed radial engine and large fixed tail-dragging gear (the** somewhat similar T-6, page 56, is a retractable tail dragger). **Tall, narrow tail fin.**

Vultee developed the basic trainer BT-13 before merging with Consolidated and built the plane through WWII; the BT-13s were still in military service as late as 1950. Known to a generation of pilots as "the Vibrator" (more a reference to what it did to airport windows than what it did to the pilots). Of the thousands that went on the war-surplus market, most were cannibalized: The Valiant's Wasp Junior radial engine fit the Stearman Kaydet, a popular sportster and crop-duster.

22

Ryan PT-21

Ryan ST

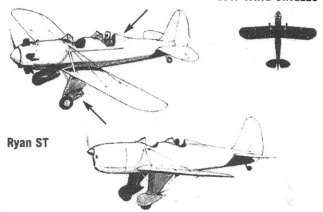

Fairchild PT-19B Cornell

Fairchild PT-19

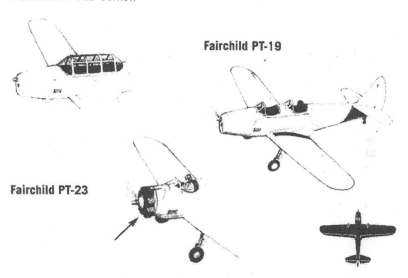

Fairchild PT-23

Consolidated Vultee Valiant, BT-13

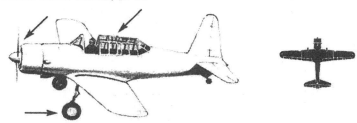

de Havilland DHC1 Chipmunk
Length: 25'5" (7.75 m) **Wingspan:** 34'4" (10.46 m) **Cruising speed:** 124 mph (200 km/h)

Rare in the United States; more common in Canada. **Unbraced low wing; fixed tail-dragging gear.** Compared to the Fairchild PT-19 Cornell, the Chipmunk has a **short, two-pane greenhouse canopy** that sits much farther back than the Fairchild's. A large air intake sits under the propeller spinner and is offset sharply to the port side of the aircraft.

Created in Canada to replace the biplane DH82 Tiger Moth as a primary trainer, the Chipmunk was built from 1946 to 1953 in Canada and Great Britain and is the most antique looking of all the post-WWII all-metal-construction aircraft. If you have a chance to see one near a Gipsy Moth or a Tiger Moth, note the similarity in the slimness of the fuselage and the shape of the engine cowling: The Chipmunk is very much a one-winged Moth.

Varga Kachina, Morrisey 2000
Length: 21'2" (6.45 m) **Wingspan:** 30' (9.14 m) **Cruising speed:** 127 mph (204 km/h)

A small, **low-wing single** of modern all-metal construction but with an **old-fashioned-looking "fighter" cockpit canopy that covers tandem seating; near constant-chord (width) wings with rounded tips; upright tail fin.**

Known as the Morrisey Nifty, a design created in wood and fabric construction by William Morrisey, a Douglas test pilot, after WWII. Redesigned in all metal in the 1960s. Many sold with **tail-dragging** gear to appeal to the owner wanting to increase the illusion of flying a WWII fighter plane. Built standard with dual controls; a popular sport and training aircraft, particularly for the weekend rental market. Morrisey has reacquired the design.

Tiger and earlier versions: Gulfstream American Yankee, AA-1, AA-5 Cheetah, Lynx, T-Cat, and AG5B Tiger
Tiger Models data: Length: 22' (6.70 m) **Wingspan:** 31'6" (9.6 m)
Cruising speed: 150 mph (241 km/h)

A series of similarly shaped aircraft that grew slightly larger over the years: **unbraced low wing, fixed tricycle gear, and constant-chord wings and tail planes. Earliest models with bubble canopy followed by a small side window (upper sketch); later models with conventional canopy (lower sketch).**

Built since 1972 with the innovative structure of unriveted aluminum skin bonded to an aluminum honeycomb. Designer Jim Bede built them under his own name, followed by American Aviation; bought by Grumman (Grumman American), then as Gulfstream American, then American General, and now simply Tiger Aircraft of Martinsburg, West Virginia.

de Havilland DHC1 Chipmunk

Varga Kachina, Morrisey 2000

Grumman American Lynx

Gulfstream American Cheetah

Beech Skipper 77
Length: 24' (7.32 m) **Wingspan:** 30' (9.14 m) **Cruising speed:** 112 mph (180 km/h)

Uncommon fixed-gear trainer. Compare with Piper Tomahawk (next entry) before deciding. Skipper has **Hershey bar wing** (with fillet fairing to leading edge) **and tail plane, true T-tail; trapezoidal side window in each door;** shorter and wider wings than the Piper Tomahawk. Skipper **main landing gear is spraddle legged,** leaning back and out, giving the plane a very wide stance on the runway.

In use by 1979, a year after the competitive Tomahawk. The primary trainer for company-franchised Beech Aero Centers. Originally planned as a conventional-tail aircraft and so flown as a prototype in 1978; the T-tail was apparently triggered by the success of the Tomahawk in 1978.

Piper PA38 Tomahawk
Length: 23'1" (7.03 m) **Wingspan:** 34' (10.36 m) **Cruising speed:** 114 mph (183 km/h)

Very common trainer. **Pure Hershey bar wing and tail plane without any fillets or fairings.** Wing is visibly longer and slimmer than on comparable Beech Skipper; **not quite a T-tail** (a cross-tail); **rectangular window in each door.**

Piper's very successful entrant into the modern trainer market, with more than 1,000 ordered in the first year (1978). Achieves the same wide stance as the Skipper (for better runway control) but without the spraddle-legged look. Tomahawk's 4'9" wheelbase was achieved by wing-mounting the main gear; Skipper's 5'2" wheelbase requires longer wheel struts, as it arises at the root of the wing and fuselage.

Ercoupe (Alon Aircoupe, Mooney M10 Cadet)
Length: 20'9" (6.32 m) **Wingspan:** 30' (9.14 m) **Cruising speed:** 110 mph (177 km/h)

Increasingly rare. **Distinctive twin fin tail is unique on single-engine aircraft; strong dihedral in constant-chord** (width) **wings; rounded wingtips.**

Designed and first built just before WWII, the Ercoupe was intended as a plane for Sunday drivers and survived until 1970 (Mooney M10 Cadet). Used a conventional steering wheel that moved the ailerons and rudder simultaneously for turning; angle of climb and descent governed normally, by pushing or pulling on the "steering column" stick. It's designed to be spin- and stall-proof, if not idiot-proof. Ercoupe also introduced the tricycle landing gear to the private pilot, making the plane astonishingly easy to fly off the runway. The lack of foot pedals made flying accessible to many handicapped pilots. (It looked so easy that the author's father talked of buying one — until the author's mother overheard him.)

Beech Skipper 77

Piper PA38 Tomahawk

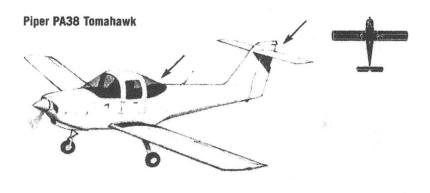

Ercoupe

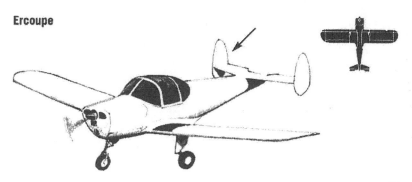

Beech Sierra (retractable), Sundowner, Sport, Musketeer
Length: 25'9" (7.85 m) **Wingspan:** 32'9" (9.98 m) **Cruising speed:** 158 mph (254 km/h)

All models quite common. The Sierra's (top sketch) **retractable gear** folds outward; wheels remain visible under wing; long, thin, rectangular tail plane; perfectly rectangular wings enter fuselage without any fairing. When you have other similarly sized airplanes to compare it with, a distinct field mark is the **high cockpit ceiling.** All two-window versions seat three; those with three or four side windows seat five, including the pilot.

Developed in 1969 as a retractable-gear Musketeer; marketed after 1970 as the Sierra. Early versions were regarded as slow and klutzy. Major changes included increased engine power (from 170 hp to 200 hp) and aerodynamic underwing fairings to shield the retracted wheels—the so-called speed bumps. Still not a high-performance aircraft, but it's roomy inside, with unusually good pilot visibility.

The Musketeer II (middle sketch): Wings and tail surfaces are identical to Sierra's but has fixed gear. Oldest models of Musketeer have two side windows.

The Sundowner (lower sketch): Distinguish it from other fixed-gear Musketeer types by the larger side windows (note rear window in particular) and the longer propeller spinner and slightly more streamlined engine cowling. A two-window Sundowner is the Sport.

Aerospatiale (SOCATA) Rallye
Length: 23'9" (7.24 m) **Wingspan:** 31'6" (9.61 m) **Cruising speed:** 108 mph (174 km/h)

Rare low-wing plane with fixed tricycle gear; large one-piece side window on glass canopy; wing and tail plane are constant-chord. When in view, note the substantial **bullet-shaped "close-out"** fairing at the tail end of the fuselage.

A variable series of small planes with two-, three-, and four-seat versions, built in France since 1958. Various names for different models: Sport, Tourisme, Club, and Minerva. It's been imported into the United States and Canada since 1974; the most common model is the 225 hp Minerva. The Hershey bar wing and tail plane resemble certain Piper models; curiously, Piper was the U.S. importer in the 1970s.

Mudry C.A.P. 10
Length: 23'6" (7.16 m) **Wingspan:** 26'5" (8.06 m) **Cruising speed:** 155 mph (250 km/h)

Distinctive little **Spitfire-shaped** plane with an apparently **oversized bubble canopy.** Rare in the United States but flown by some flight-instruction programs and thus common locally.

The Mudry is probably the best example of the decline and fall of the U.S. light-plane industry in the face of high premiums for product liability insurance. No one was making a little, inexpensive, side-by-side aerobatic plane for advanced **civilian** pilot training, so this French import (first flown in 1970), based on the old home-built Piel Emeraude, was imported in the 1990s.

Beech Sierra

Musketeer II

Sundowner

Aerospatiale (SOCATA) Rallye

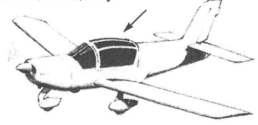

Mudry C.A.P. 10

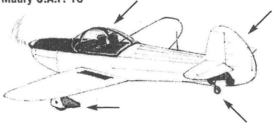

29

Liberty XL
Length: 20'6" (6.28 m) **Wingspan:** 28' (8.53 m) **Cruising speed:** 150 mph (245 km/h)

Adapted from the Europa series of touring and training planes, the XL has a plenitude of glass: **fixed tricycle gear with extensive wheel pants, three noticeable flap guides under each wing, wings set high in the fuselage, underspinner air scoop, gigantic windows set in gullwing doors.**

The Liberty XLs first flew in the mid-1990s, built as a two-person touring plane with the range and speed and comfort associated with four-seaters: a sort of airplane equivalent of the popular German two-seat roadsters. The wings insert into "wing boxes" set in the airplane's steel frame and are held there by motor-driven pins. It is relatively simple to unpin the wings, put them on a trailer along with the rest of the plane, and take it home and put it in the garage. Versions of this technique are expressed in dozens of models of sailplanes.

Grob G120A
Length: 26'5" (8.06 m) **Wingspan:** 33'5" (10.2 m) **Cruising speed:** 190 mph (307 km/h)

A chunky low-wing with lots of appendages: **two typical trainer features—outsized canopy and pantless wheels—a small dorsal finlet just behind the canopy, and noticeable strakes slightly forward of the tail.**

Extremely uncommon so far but may be noticeable in the Phoenix, Arizona, area, where Lufthansa uses it in pilot training (with the stylized eagle emblem on the tail fin). Fully aerobatic and highly maneuverable at all speeds.

Diamond DA40 TDI Diamond Star and Diamond DA20 Katana
DA40 TDI data: Length: 26'3" (8.01 m) **Wingspan:** 37'2" (11.94 m)
Cruising speed: 125 mph (244 km/h)

A distinctive family of wasp-waisted aircraft developed from German motorgliders: **high-aspect (long and narrow) low wings with soft-curved winglets; abrupt compression of the fuselage into a tail boom from the bulbous cabin to the T-tail. Some DA20s have wheel pants, but most trainers do not.**

Popular trainers, the slightly quicker bubble-canopied side-by-side two-seat DA20s are in the air many more hours than the newer, four-passenger, more sedate DA40s. Since 2002, a version of the DA20 with stronger landing gear and a higher bubble has been the primary flight trainer at the U.S. Air Force Academy where a private contractor provides the training. At the end of 2004, the planes had not yet been given a military designation. The composite materials allow for the long and narrow fuselage, a common sight at any sailplane meet from several manufacturers but unusual in general aviation. The DA40 TDI Diamond Star boasts being the first turbodiesel in production, with diesel's generally higher mileage and lower cost transferred from the automobile to the airplane. The diesel promises to sell well in Europe.

Liberty XL-2

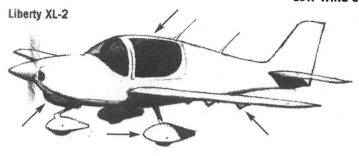

Grob G120A

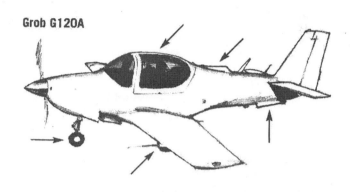

Diamond Star DA40

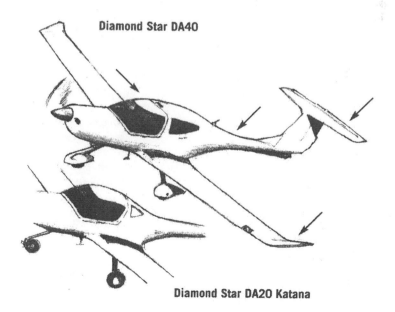

Diamond Star DA2O Katana

Lancair Columbia 300, 350, 400
Length: 25'2" (7.68 m) **Wingspan:** 36' (10.97 m) **Cruising speed:** 220 mph (352 km/h)

Resembles other modern composite-material aircraft; see the Cirrus Design Aircraft (next entry). **Perches atop its one-piece wing; cabin drops down the windscreen giving a bit of a hunched look; extremely swept-back tail fin; note the forward-thrusting strut to the fixed nose gear.**

Columbias are built from carbon-fiber materials (similar to the graphite fishing rods that dominate the industry). The high cruising speed seems impossible for a plane with fixed gear, but superb aerodynamics and a turbocharged engine in the 400 make it truly happen.

Cirrus Design Aircraft
Length: 26' (7.92 m) **Wingspan:** 35'8" (10.88 m)
Cruising speed: for the middle-range version, 180 mph (290 km/h)

Externally identical, all Cirrus models have a **small skylight window behind the passenger cabin, curved miniwinglets, elaborately curved wheel pants, a swooping leading edge on the tail fin, a pair of prominent air scoops, and, visible on the ground, automobile-style doors.**

Cirrus adopted the first system for lowering an engine-failed airplane to the ground: a pilot-controlled, three-point harness ballistic parachute from BRS Parachutes. A few other planes have been retrofitted (Cessna 150s, mostly), but Cirrus designed the parachute into the aircraft's fuselage from day one. The chute, located behind the skylight window, is covered with a tearable skin, and the two harnesses that are attached up by the windscreen lie in grooves covered with the same material.

Slingsby Firefly T-3A (United States), CT-111 (Canada)
Length: 24'9" (7.5 m) **Wingspan:** 34'9" (10.6 m) **Cruising speed:** 155 mph (250 km/h)

Characterized by a **long and thin wing, a quickly tapering fuselage** (but not truly wasp-waisted), side-by-side seating covered by a **substantial bubble, and three air scoops.** Always with **fixed gear,** usually without wheel pants.

This popular trainer and club plane has had a curious history in the United States, with three fatal accidents and numerous incidents when it was the primary trainer at the U.S. Air Force Academy in Colorado. Grounded by the military in the United States, the plane continues to be Canada's primary trainer, with no untoward problems. In both countries, the initial military training is done by private contractors. The fatal accidents with the T-3 were with U.S. Air Force trainers aboard. It has been replaced as of this writing by a modified (rough-landing-resistant) Diamond DA20 (page 30).

Lancair Columbia 300, 350, 400

Cirrus Design Aircraft

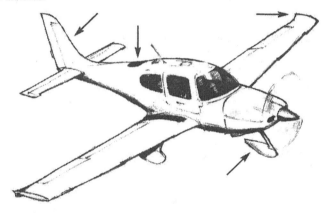

Slingsby Firefly T-3A (U.S.)/ CT-111 (Canada)

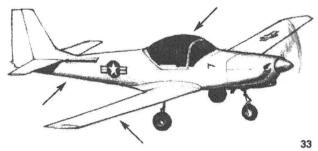

Piper PA28 Cherokee 140, 150, 160, Charger, Flite-Liner
Length: 23'3" (7.08 m) **Wingspan:** 30' (9.14 m)
Cruising speed: with 180 hp engine, 130 mph (209 km/h)

Common. Small four-seater, Hershey bar wing, fixed tricycle gear with wheel pants.

Introduced in 1961; superseded by the Cherokee Warrior in 1974, when it received the multiangled "new Piper" wing. Engines built with 140 hp to 235 hp. The plane was eventually designated Charger. When stretched to hold six, it became the Cherokee SIX (page 38). The 150 hp version, designated Flite-Liner, was a popular club plane and trainer in the 1970s. The original Cherokee introduced to the small-plane market considerable use of simple curves and fiberglass and plastic construction.

Piper PA28 Cherokee Warrior, Warrior II, Cadet, Dakota, Archer II
Length: 23'9" (7.25 m) **Wingspan:** 35' (10.67 m) **Cruising speed:** 130 mph (209 km/h)

Common. Fixed tricycle gear; dihedral in wing, none in tail; three side windows. Wing is of complicated geometry: leaves fuselage with fairing to leading edge; short equal-span section; leading and trailing edges taper to tip at unequal angles. Tail plane a pure Hershey bar rectangle.

Flown since 1974, the first Piper to abandon its trademark of constant-chord wing plans. Sold under various names with slight differences, including engine horsepower: Cherokee Warrior, renamed Warrior II (160 hp), Dakota (235 hp), Archer II (180 hp). All versions seat four, including the pilot. Cadet, a trainer, drops third side window and, like most bouncing trainers, the wheel pants.

New Piper Archer III, New Piper Warrior III
Length: 24' (7.32 m) **Wingspan:** 35'6" (10.8 m) **Cruising speed:** 135 mph (217 km/h)

A family of aircraft started in 1974 with the change from the old Piper Hershey bar wing to a modern tapered design; noticeable dihedral in wings contrasts with none in the tail plane; wing and tail identical to the PA28 Warrior (previous entry).

The New Piper Archer III is a reprise of the Archer II but can be distinguished by the new cowling with a raised section from the spinner to the windscreen, and two circular air intakes. See sketch.

If you see what appears to be an "old" Piper PA28 without wheel pants, it is (if truly old) the dual-control trainer, the Cadet. Training aircraft get bounced too hard to keep their pants on. If you see a shiny new Piper PA28 without pants but with fairings enclosing the main wheel struts and without the new Archer III cowling, it is the New Piper Warrior III dual-control trainer.

Piper PA28 Cherokee 150

Piper PA28 Warrior

New Piper Archer III

Piper PA28-180R Cherokee Arrow, Arrow II, Arrow III
Length: 24'2" (7.37 m) **Wingspan:** 32' (9.75 m) **Cruising speed:** 162 mph (261 km/h)

Less common than the nonretractable Cherokee series. **Identical to the fixed-gear Cherokees** (see the two entries for Piper PA28 Cherokee and Cherokee Warrior field notes). For simplicity's sake: The Arrow II (illustrated) has three side windows and constant-chord wings; a two-window Arrow is a I. The Arrow III has the tapered Piper wing and is identical to the Cherokee Warrior II, with tapered wings, except for its retractable gear. There are a few Arrow IIIs with turbocharged engines (see sketch for Turbo Arrow IV, page 37, showing the turbocharger air scoop). On the flight line with wheels down, an Arrow is a **Cherokee without wheel pants.** On the air traffic controller's radio, they're all just plain Cherokees.

Piper PA28RT Arrow IV
Length: 27' (8.23 m) **Wingspan:** 35'5" (10.8 m) **Cruising speed:** 165 mph (265 km/h)

Not especially common. What we have here is **a Cherokee Warrior II with a T-tail.** Has the **tapered wings of the Warrior series** (page 34). A much larger plane than the little T-tailed Beech Skipper; fully retractable gear.

If there was ever any proof that the T-tail had some sales advantages, as opposed to utilitarian purpose, it was sticking one on the old, reliable Cherokee Warrior II/Archer airframes in 1977. The T-tail Arrow IV came in conventional and turbocharged models (see lower sketch, showing air intake), as did the Arrow III.

New Piper Arrow
Length: 24'8" (7.52 m) **Wingspan:** 35'5" (7.52 m) **Cruising speed:** 158 mph (254 km/h)

Neither the old T-tail Arrow IV nor the new conventional version are common. **Retractable gear, three side windows, dihedral in tapered wing; the tail plane is rectangular and without dihedral.**

This New Piper aircraft is really just the good old T-tailed plane with a new rear end. It is almost as large as, and looks like, a Saratoga that has lost one window.

Piper PA28-180R Cherokee Arrow II

Piper PA28RT-201 Arrow IV

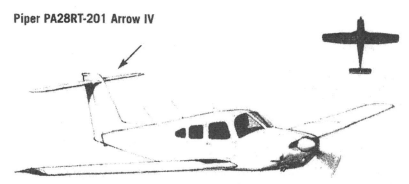

Piper PA28RT-201T Turbo Arrow IV

New Piper Arrow

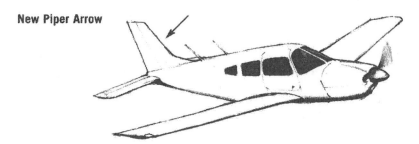

Piper PA32 Cherokee SIX, PA32R-300 Lance, PA32RT-300 Lance II
Length: 27'9" (8.45 m) **Wingspan:** 32'9" (9.95 m) **Cruising speed:** 158 mph (254 km/h)

A common, **large, fixed-gear airplane.** Typical early Piper wing: a Hershey bar rectangle with fairing to leading edge; **an oversized Cherokee with four side windows.** The earliest models had four squared windows, not the variable geometrical shapes seen in the sketch. A retractable Cherokee SIX, with Hershey bar wings, is a Lance, of which a few models had T-tails (lower sketch).

Carrying six, including the pilot, it was Piper's largest single-engine and the largest fixed-gear single in the private-aviation field for many years (1964–1979). When equipped with an optional 300 hp engine, it's suitable for use on skis or floats. Occasionally used as an air ambulance or short-haul freighter; then equipped with a single large door that folds up at the rear of the cabin. Last produced in 1979, when Piper replaced it with the nonretractable PA32 Saratoga, using the longer, tapered, "new Piper" wing plan.

Piper PA32R-301 Saratoga
Length: 28'4" (8.64 m) **Wingspan:** 36'2" (11.02 m) **Cruising speed:** 162 mph (261 km/h)

What we have here is a **Cherokee SIX with the new, tapered Piper wing (upper sketch).** If you can't get a look at the wing, call it a Cherokee.

The Saratoga is a six-passenger addition, usually sold with retractable gear, many with turbocharged engines (see lower sketch). First produced in 1979, the Saratoga replaced the Cherokee SIX and the T-tailed Lance. The name change signifies mostly the wing change, as well as more horsepower.

PA32R, New Piper PA32 Saratoga
Length: 27'9" (8.5 m) **Wingspan:** 36'2" (11 m) **Cruising speed:** 200 mph (324 km/h)

The New Piper Saratoga is **always with retractable gear and always with the same new cowling (see sketch), which may be covering a conventional or a turbocharged engine.**

The original Saratoga was usually built with **retractable gear; the few with fixed gear look like a Cherokee SIX with a new, tapered wing. When the older Saratoga is supercharged, it has an underspinner air intake (see sketch).**

The retractable "old Piper" Saratoga quickly replaced the other six-passenger aircraft in the Piper line (the Lance II and the Cherokee SIX). The New Piper Saratoga has much more advanced (and expensive) avionics and considerably more speed with either type of engine.

New Piper Cherokee 6X and XT
Length: 27'9" (8.5 m) **Wingspan:** 36'2" (11 m) **Cruising speed:** 185 mph (298 km/h)

Although dubbed a Cherokee, it's a **Saratoga with fixed gear and wheel pants, one less window, and the New Piper cowling, which may conceal a conventional or a turbocharged engine.**

Piper PA32RT-300 Lance II

Piper PA32 Cherokee SIX

Piper PA32R-301T Turbo Saratoga

Piper PA32R-301 Saratoga

New Piper PA32 Saratoga

New Piper Cherokee 6X

Beech Bonanza 35, F33A

Length: 26'5" (8.05 m) **Wingspan:** 33'6" (10.21 m) **Cruising speed:** 190 mph (306 km/h)

Anything with a V-tail is a Bonanza 35. Confusion is generated by two conventional-tail aircraft: the Bonanza A36 (next entry) and the Bonanza 33, which is identical to the Bonanza 35 except for its conventional tail. (See the next entry for details.)

Built from 1947 to date, more than 10,000 are flying in North America. Before 1961, about 1,200 were built with only two side windows; however, some owners have added the third side window to their own pre-1961 aircraft. It comes with a variety of engines, including turbocharging. Early models had a smaller tail surface, less steeply angled, but after-market modifications have been made to most of those models. Of all-metal construction since its inception.

Beech Bonanza A36

Length: 27'6" (8.38 m) **Wingspan:** 33'6" (10.21 m) **Cruising speed:** 188 mph (302 km/h)

Commonest of the large, single-engine, retractable-gear planes. **Fairing from fuselage to wing's leading edge; four side windows; large doors on starboard side.** If you put a Beech 36 conventional tail on the Beech 35 (preceding entry), you have the Beech Bonanza 33 (once known as the Debonair).

Built since 1968, the Beech 36 seats six, including the pilot; for many years, the only six-passenger, retractable-gear single. Turbocharged model (illustrated) shows intake and cooling louvers on engine cowling. The smaller Debonair/Bonanza 33 has three side windows and seats four, including the pilot. Since 1982, the turbocharged model has a 37'6" (11.43 m) wingspan. A few turboprop conversions, with wingtip tanks, have been made.

North American Rockwell Commander 111, 112, 114

Length: 25' (7.62 m) **Wingspan:** 32'11" (10.04 m) **Cruising speed:** 157 mph (253 km/h)

Not common. Best field mark for this low-wing single is the **tail plane, mounted midway up the tail fin.** Overhead, the **wing leading edge is straight, at right angles to the centerline,** except for the noticeable fairing from fuselage to leading edge. Strong (7-degree) dihedral in wing, none in tail plane. A wide, chubby look to the cabin area.

Built since 1971, it's a high-performance, four-seat single. The unusual tail design caused some difficulty at first, including the loss of a prototype, and the requirement to redesign the rear fuselage and tail assembly. The interior cabin space is unusually wide for a four-passenger single and gives the aircraft its look of being bulky forward and over the wing. With a three-bladed prop, it's a Commander Aircraft 114B.

Beech Bonanza 35, F33A

Beech Bonanza A36

North American Rockwell Commander 112

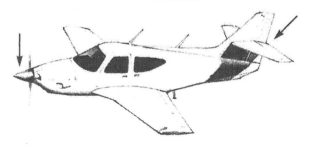

EADS TBM 700

Length: 34'2" (10.41 m) **Wingspan:** 39'11" (12.16 m) **Cruising speed:** 345 mph (555 km/h)

Longer, slimmer, shorter span than the Piper Malibu (next entry), which it somewhat resembles. **Long nosed, with air scoop below small four-bladed propeller, engine exhaust visible starboard, noticeable dihedral in horizontal stabilizer, odd "bent-down" pilot's side window.**

A successful 1989 introduction, as American manufacturers concentrated on corporate pure jets, Mooney and SOCATA (France) combined to produce this pressurized 30,000 ft.-ceiling turboprop business aircraft. Its cruising speed and rate of climb, 2,303 ft. (702 m) per minute, put it nearly in jet performance; its low stall speed, 71 mph (113 km/h), makes it amenable to small airports and noise-restricted areas.

New Piper Malibu Mirage and Malibu Meridian, Piper PA46 Malibu

Length: 28'9" (8.8 m) **Wingspan:** 43' (13.1 m) **Cruising speed:** 245 mph (394 km/h)

These Malibu Pipers, new and old, are unlike all other Pipers: **a rounded "heavy" cabin, paired with extremely long and narrow wings (high-aspect wings).**

The original Malibu was always turbocharged. The rounded fuselage makes for a stronger "pressure vessel," allowing a service ceiling of 30,000 ft. (915 m). In the New Piper versions, the turbocharged Meridian shows large exhausts lined up with the spinner. The conventionally aspirated Mirage has the New Piper cowling without side-venting exhausts. (See sketch.) Even the conventional new Malibu Mirage can scorch an "Old Piper" Malibu by 62 mph (100 km/h) at maximum cruising speed.

Pilatus PC-12

Length: 47'3" (14.4 m) **Wingspan:** 53'3" (16.23 m) **Cruising speed:** 310 mph (500 km/h)

The largest single-engine plane you'll see: **T-tail with a bullet (like a Beech 1900) five windows; long, thin, low wings with oversize winglets (right winglet begins at a radar dome).** The high-aspect wings seem too thin for the bulky fuselage, but it flies.

This craft, Swiss designed and built, can compete against business jets; though slower, its long range—more than 2,500 mi., or 4,000 km—gives it near parity with business jets that have to refuel more frequently. Its long range and good "hang time" give it a role in surveillance and search and rescue. Although most countries prohibit single-engine passenger certification, it is flown in Canada as a nine-passenger mini airliner.

EADS TBM 700

Piper PA46 Malibu

Pilatus PC-12

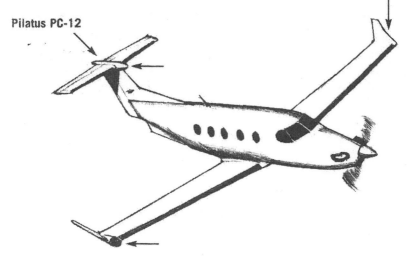

MOONEY FOUR-PASSENGER AIRCRAFT

A complex group of airplanes that grew out of the original four-place model M20. The only Mooney with a different basic model number is the M2 Mustang, a very rare pressurized aircraft. All have **tricycle gear (almost always retractable) and have a set of common field marks: All leading edges—wing, tail fin, and tail plane—are straight lines perpendicular to the centerline of the aircraft. The trailing edges are slightly but noticeably angled forward,** giving the airplane an appealing look of eagerness. The original Mooney M18 Mite (page 54) was a small tail dragger but with the same wing and tail format.

Mooney M20, 201MSE, 205, 231, 252, PFM

201MSE data: Length: 24'8" (7.52 m) **Wingspan:** 36'1" (10.67 m) **Cruising speed:** 167 mph (269 km/h)

Last model had rounded-off side windows; earlier 201, 231, 205 had square-edged windows. Porsche-engined high-performance PFM (see sketch) and a turboprop 231 have longer, sleeker engine cowlings; 205 has fully enclosed landing gear.

Mooney M20F, M20J (in part), M20 PFM

Typical M20F data: Length: 24'8" (7.52 m) **Wingspan:** 36'1" (10.67 m)
Cruising speed: 167 mph (269 km/h)

More than 3,000 of the "long" Mooneys with the larger, squared-off windows were built before the introduction of the rounded windows (previous entry) and were sold as models 201 or 231. Similar aircraft with longer cowlings include the Porsche-powered M20 PFM and a turbocharged M20 231.

Mooney M20D (in part), M20E "Chapparal" and variants

Length: 23'2" (7.06 m) **Wingspan:** 35' (10.67 m) **Cruising speed:** 172 mph (227 km/h)

The last of these "short" Mooneys were built in 1998, all characterized by the angle-edged side windows and a one-piece windshield (compare with the M20C and some M20Ds). When the company was owned by the Aerostar Corporation (1969 to 1971), most M20Es were given a meaningless but interesting bullet on the top of the tail fin (see sketch) and marketed as Mooney Aerostars.

Mooney M20

Mooney M20 201MSE

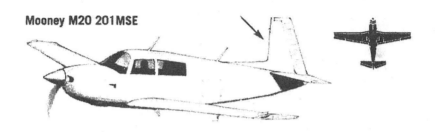

Mooney M20F, J

M20 PFM

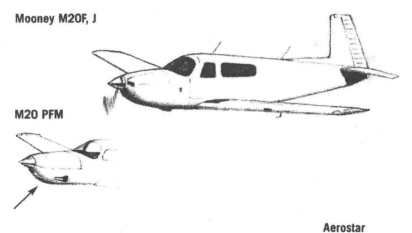

Aerostar

Mooney M20 Chapparal

Mooney M22 Mustang
Length: 26'10" (8.18 m) **Wingspan:** 35' (10.67 m) **Cruising speed:** 214 mph (344 km/h)

Rare, built only from 1967 to 1969. Pressurized, which shows in the window design: four small side windows (three square and a round trailing window). A very high-performance single, with a 24,000 ft. operating ceiling.

Mooney M20C, M20D Master
Length: 23'2" (7.06 m) **Wingspan:** 35' (10.67 m)
Cruising speeds: 130 mph to 150 mph (209 km/h to 241 km/h)

The original metal-clad Mooney model 20s, unlike their descendants, have a peculiar small, droopy rear-passenger window and a two-piece windscreen. The Mooney Master was a brief experiment with fixed tricycle gear.

Mooney sold nearly a thousand fabric-covered airplanes (models M20, M20A, and M20B) that will be indistinguishable at any distance from the model M20C—all-metal models that sold more than 2,000 copies before merging onward to the Mooney M20F (page 44). The rather rare fixed-gear M20D Master sold only 160 aircraft before the idea was abandoned.

EADS SOCATA TB 9 Tampico, TB 10GT Tobago, TB 20 Trinidad or Tobago, TB 20GT Trinidad, TB 21GT Trinidad
TB 21GT data: Length: 25'5" (7.75 m) **Wingspan:** 32'9" (9.97 m) **Cruising speed:** 218 mph (352 km/h)

A family of very similar-looking aircraft: The TB 20 and 21 have retractable landing gear, a noticeable strake, and a small winglet on the rear half of the wing. In most light and at any distance, the side windows appear to be a single piece.

Later models 9, 10, and 200 have fixed gear and wheel pants that taper into a vertical stabilizer. The model numbers reflect different engine options. The original Tampico (TB 9) is an extremely popular shared-ownership, "club" aircraft.

The addition of the GT designation to previously used names and model numbers reflects a redesigned cabin with flush-mounted windows on all models, fancy pants on all the fixed gears, and strakes on the newest retractable 20 and 21GTs.

Mooney M22 Mustang

Mooney M20D Master

Mooney M20C

EADS SOCATA TB 20/21 Trinidad

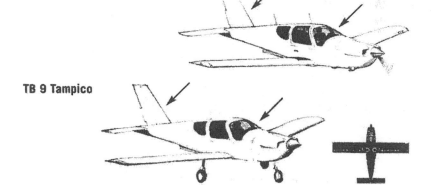

TB 9 Tampico

EADS SOCATA Trinidad TB 21GT

Aeromacci SIAI-Marchetti SF.260

Length: 24'4" (7.1 m) **Wingspan:** 27'5" (8.25 m) **Cruising speed:** 200 mph (322 km/h)

The combination of a bubble canopy and wingtip tanks is unique; the fairing to the tail fin begins just behind the canopy.

This basic design comes in several variations, including one with the stronger landing gear for military trainers and counterinsurgency uses. Used in several countries, including Mexico, for the second stage of pilot training. Popular with businesses offering "fighter pilot" experiences.

SIAI-Marchetti S.205, S.208

S.205 data: Length: 26'3" (8 m) **Wingspan:** 35'7" (10.86 m)
Cruising speed: 140 mph (226 km/h)

One of the few singles with a typical cabin and wingtip tanks (see Navion, page 50), but not all models delivered had them! Most U.S.-based planes have retractable gear; the nose wheel remains visible. Perhaps the best field mark is the very upright tail; forward edge leans back; trailing edge is vertical; distinct dorsal fairing to tail. It looks like a Mooney tail fin installed backward.

Extremely variable, these SIAI 205s and 208s, with horsepower from 180 (S.205-18) to 260 (S.208) and cruising speeds from pokey to fast. Most carry a pilot and three passengers.

Aero Commander 200 (Meyers 200)

Length: 24'4" (7.42 m) **Wingspan:** 30'6" (9.29 m) **Cruising speed:** 215 mph (346 km/h)

Quite rare. A small, retractable tricycle gear, distinguished by a high cabin canopy and an automobile-type door on starboard side of the cabin. Aft of bulbous canopy, appearance is short winged, slim fuselaged. Could be confused with the Ryan Navion (next entry).

Aero Commander took over the Meyers 200, buying a design that put it in the high-performance, four-seat, retractable market in 1965. Very few Meyers 200s and not many more (perhaps 100) Aero Commander 200s were built from 1965 to 1967. Built with various engines, including one type with a turboprop, the Interceptor 400, with cruising speeds near 300 mph (400 km/h). More fun than practical to fly.

Aeromacci SIAI-Marchetti SF.260

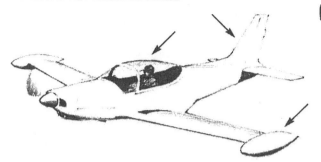

SIAI-Marchetti S.205, S.208

Aero Commander 200

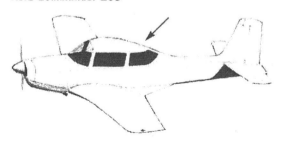

Ryan Navion (L-17), North American Navion

Length: 27'8" (8.43 m) **Wingspan:** 33'5" (10.18 m) **Cruising speed:** 155 mph (249 km/h)

Rare. A low-wing single with a **bulbous cockpit canopy and slender rear fuselage. Nose wheel is visible** when tricycle gear is retracted. Could easily be confused with the even rarer Aero Commander 200 (previous entry): **Navion's rear side window tapers sharply; two-piece windshield has noticeable center strip,** whereas the Aero Commander has a much larger rear window that sweeps up and a one-piece windshield.

Manufactured in the late 1940s through 1951, it seats four, including the pilot. Ryan built hundreds of low-wing trainers during WWII but purchased the Navion design from North American. Came standard with dual controls and a bench seat for two more passengers. Canopy slides back for access to cabin. Ryan added landing gear doors and personal-comfort items to the basic North American design.

Navion Rangemaster

Length: 27'6" (8.38 m) **Wingspan:** 34'9" (10.59 m) **Cruising speed:** 290 mph (467 km/h)

A rare, odd bird: a **low-wing single with wingtip tanks.** It's essentially similar in wing and tail configuration to the Ryan Navion (previous entry), but with a built-up five-passenger cabin and automobile-type door on the port side of the aircraft.

A Texas aircraft parts manufacturer picked up the old Ryan Navion design, spare parts, and tools to manufacture the Rangemaster—all quite similar except for the cabin—and supplied with a variety of engines. Like the prototype, it comes standard with dual controls.

Piper PA24 Comanche

Length: 25' (7.62 m) **Wingspan:** 36' (10.98 m) **Cruising speed:** 182 mph (293 km/h)

Chunky fuselage. Commonly, two side windows; last models had three. Retractable gear is visible, tucked in against fuselage. Beech Bonanza–type wing, fairing to a straight leading edge, tapered trailing edge.

Piper's first low-wing was also its first retractable. The wing appears to be a Beech borrow but is in fact a U.S. government design with several thousand built before production ended in 1972. In the last few years of production, the plane stretched the cabin to seat five or six and added the third window, at which point Piper shifted to the Arrow series (page 36) as the standard six-passenger retractable.

Ryan Navion (L-17)

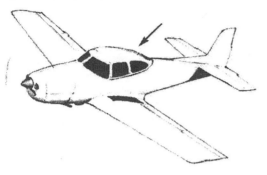

Navion Rangemaster

Piper PA24 Comanche

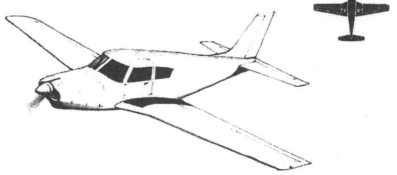

Temco (Globe) Swift 125
Length: 20'11" (6.38 m) **Wingspan:** 29'4" (8.94 m) **Cruising speed:** 140 mph (225 km/h)

Rare. A small, retractable, low-wing; cockpit and windows varied, not good field marks; strong (8-degree) dihedral in tail plane and wings— very unusual in small singles and a distinct field mark at any altitude or attitude. Close at hand, a unique engine grill, like something from a 1950s General Motors automobile.

A few hundred of these 1945–1951 airplanes survive. They came standard with dual controls, some with all-Plexiglas canopy, some with enclosed cabin. Along with the Mooney Mite, one of the first post-WWII airplanes to take advantage of the wind tunnel-tested wing designs of the National Advisory Committee on Aeronautics (NACA), precursor of NASA. Many fly today with much more powerful engines than the original 125 hp.

Bellanca Viking (and Cruisemaster 14193C)
Length: 26'4" (8.02 m) **Wingspan:** 34'2" (10.41 m) **Cruising speed:** 185 mph (298 km/h)

A small low-wing; large, strongly swept tail fin; strut under tail planes; dihedral in wing, none in tail plane; wraparound windshield; two large side windows; nose wheel does not retract fully, main gear carried in underwing fairings.

Bellanca took the Cruisemaster (next entry), added a tricycle gear, and dropped the outboard fins on the tail planes to make the Cruisemaster 14193. The swept tail fin was added in 1958; the name changed to Viking in 1966. No longer manufactured, although efforts are occasionally made to reintroduce it. Constructed of fabric over plywood and tubing.

Bellanca Cruisemaster, Cruiseair
Cruisemaster data: Length: 22'11" (7 m) **Wingspan:** 34'2" (10.41 m) **Cruising speed:** 180 mph (290 km/h)

Rare. A stubby, low-wing tail dragger; main gear remains exposed when retracted; triple-tailed, with central tail fin much larger than outboard fins; wire braces on tail plane; two side windows.

About 100 Cruisemasters and a few hundred very similar Cruiseairs (smaller engines) were built from 1946 to 1958. Plane combined relatively high operating speeds with low landing speeds and a stall speed of about 50 mph (80 km/h). Highly regarded for sport use. Seats three or four, including the pilot. Construction is fabric over plywood.

Temco Swift 125

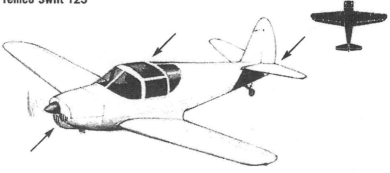

Bellanca Viking

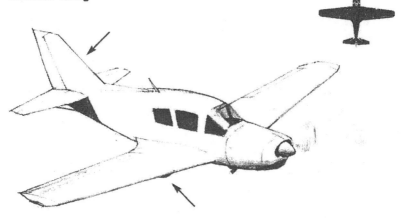

Bellanca Cruisemaster

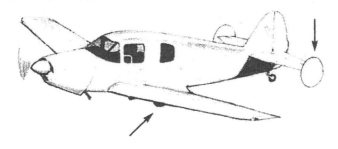

Mooney M18 Mite
Length: 18' (5.48 m) **Wingspan:** 26'10" (8.2 m) **Cruising speed:** 80 mph (129 km/h)

Rare. A classic Mooney design. Though tiny, a one-seater, it has same wing and tail-surface pattern as the four-seat Mooneys: **Leading edges of wing and tail surfaces are a straight line at right angles to the center-line of the fuselage.**

Built from 1947 to 1954, the Mooney Mite was a favorite sport plane for ex-fighter pilots—inexpensive to own and cheap to fly—but it did not answer the needs of the family-oriented pilot. Originally designed to use the old Crosley automobile engine, the last models (M18) had a regulation 65 hp aircraft engine. Still available in kit form. The first post-WWII civilian aircraft to use a NACA wing design.

Culver LCA Cadet
Length: 17'8" (5.3 m) **Wingspan:** 26'11" (8.1 m) **Cruising speed:** 120 mph (193 km/h)

Rare. A very small, low-wing retractable; dihedral in wings, none in tail plane. Overhead is a semielliptical curve to both edges of wings and tail plane. Plane has a distinct sculptured look to it, with smooth curves everywhere, as though carved from a bar of soap. Structure is mainly wood, with early fiberglass reinforcement and fuselage skin.

Built from 1939 through WWII, with a few bench-built copies as late as 1960. Final design was by Al Mooney, creator of the Mooney line of aircraft, the fastest and nimblest of pre-WWII private aircraft. During the war, used as radio-controlled target drone and pilot flown as "camera-gun" target for training Air Force gunners and pilots. So acrobatic, it was a satisfactory imitation of the hottest enemy fighter planes. It is one of the curiosities of life that Al Mooney was never brought in to design U.S. fighter planes.

Beech T-34A, B Mentor
Length: 25'10" (7.8 m) **Wingspan:** 32'10" (10 m) **Cruising speed:** 160 mph (257 km/h)

Not common. Large greenhouse canopy over tandem dual-control cockpit; large, slablike, upright tail fin. The clear "trainer look" combined with a nonradial engine separates the Mentor from the Texan and the Trojan.

In civilian hands, a popular low-wing aerobatic aircraft. In military service from 1954 to 1960 as a common U.S. Air Force and Navy basic trainer, replacing the T-6 Texan. Flown by the Navy only from 1960 to 1980. The Air Force moved to all-through jet training from 1960 to 1964, when most of the civilian-owned Mentors came on the market. Curiously, after all-through jet training was deemed a failure by the Air Force, it turned to Cessna's 172 Skyhawk (page 86), a slow, high-wing prop plane, for the first 30 hrs. of training, designating it the T-41 Mescalero.

Mooney M18 Mite

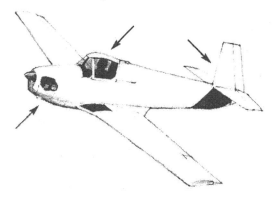

Culver LCA Cadet

Beech T-34 Mentor

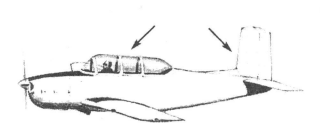

North American T-6 Texan, Harvard II
Length: 29'6" (8.99 m) **Wingspan:** 42' (12.8 m) **Cruising speed:** 218 mph (351 km/h)

A fairly common relic of WWII. **Long greenhouse canopy** over tandem dual controls; **dihedral in wing begins a few feet out from fuselage** (a reverse gullwing, as in Corsair). Close at hand or overhead, note the **rounded bump where the leading edge of the wing meets the fuselage;** this is a fairing to hold the retracted main gear wheels. **Tail fin is quite triangular.**

Built before 1941 and in service through the Korean conflict, the Texan, purchased as military surplus, was a popular sport plane for veteran pilots. More often seen parked than in the air. When flying, attention-attracting noise. More than 15,000 produced between 1941 and 1951. Overhead, the wing is typical of pre-WWII design: nearly straight trailing edge, tapering leading edge—like a single-engine DC3.

North American T-28 Trojan
Length: 32' (9.76 m) **Wingspan:** 40'1" (12.23 m) **Cruising speed:** 190 mph (306 km/h)

Not common. In civilian colors; **fat engine cowling** houses large radial engine; **long, high, Plexiglas canopy** sits atop tandem-seating dual controls; **tall, sharply angular tail fin. Plane is heavy, chunky.**

In the 1950s and the 1960s, the common U.S. armed forces basic trainer. Sank like a rock with engine failure. Like many trainers, it was adapted to a counterinsurgency role with underwing bomb and rocket mounts. A counterinsurgency role usually implies enough power to carry bombs but only against a lightly defended target. There have been a few civilian conversions, with cabins replacing the cockpit/canopy, but the general configuration is unchanged.

Grumman TBF-1 (TBM-1) Avenger, "Borate Bomber"
Length: 40' (12.2 m) **Wingspan:** 54'2" (16.5 m) **Cruising speed:** 240 mph (386 km/h)

A very rare, large, **single-engine** military aircraft. **Original greenhouse cockpit canopy usually modified but not in any standard manner; lower fuselage (bomb bay) steps up to tail section; square-cut tail surfaces.**

Now restricted to museums and air shows, except for a few that are flying, particularly in Canada, as aerial forest fire fighters, dropping "borated" or otherwise treated water on fires. Gawky, ungainly, but a fairly successful torpedo bomber. Held a crew of three: the pilot, bombardier/navigator, and gunner. The TBM-1 was identical, manufactured by General Motors under license from Grumman.

North American T-6 Texan

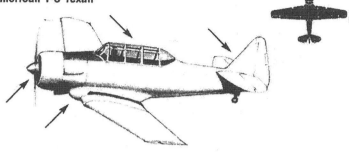

North American T-28 Trojan

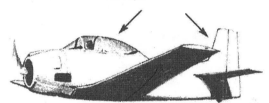

Grumman TBF-1 Avenger

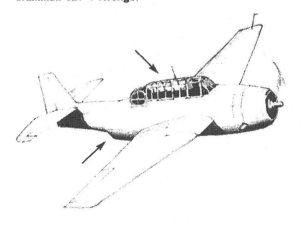

Chance Vought F-4U Corsair

Length: 33'8" (10.26 m) **Wingspan:** 41' (12.49 m) **Cruising speed:** 350 mph (563 km/h)

Unmistakable. A large, noisy, radial-engine warship with a one-man cockpit set halfway back on the fuselage. Wings drop down from fuselage, then show sharp dihedral to tip: reverse gullwing. May be seen with the wings folded up in hangars.

More than 12,000 F-4Us were produced through WWII; saw most service in 1944 and 1945. One of the most powerful fighter-bombers ever built: 2,000–3,000 hp, six .50-caliber machine guns, and 2 tn. bombs or rockets. Nicknamed Whistling Death by Japanese pilots.

North American P-51 Mustang

Length: 32'3" (9.83 m) **Wingspan:** 37' (11.28 m) **Cruising speed:** 390 mph (628 km/h)

Rare. Most often seen at air shows. Long, slim nose with massive propeller spinner. From the side or below, note that the radiator air intake for the liquid-cooled engine is set well back under the cockpit (visible in lower sketch). Tail arrangement is unusual: Tail planes set very high and well forward (to clear the full-length rudder on the tail fin).

Developed by North American in 1940 to meet a British specification for a long-range fighter-escort for British bombers that would operate over Europe from bases in England. Lower drawing shows the most common P-51D, with a bubble canopy for good vision to the rear.

The P-51A–C types have a turtleback style (upper sketch). The Cavalier Aircraft Company has built modern P-51-Ds with wingtip tanks to be used as counterinsurgency planes by U.S. allies. This design was acquired by Piper Aircraft, which continued to develop the aircraft as the Enforcer until 1984. Counterinsurgency aircraft, as we have noted, are best defined as easily maintained fighter-bombers for use against lightly defended persons and dwellings.

Curtiss P-40 Warhawk, Tomahawk, Kittyhawk

Length: 33'4" (10.05 m) **Wingspan:** 37'4" (11.3 m) **Cruising speed:** 315 mph (507 km/h)

Rare but recently increasing to a few dozen. Compared to other WWII restorations: low, rounded tail fin; huge and obvious air scoop under quite pointed propeller spinner; greenhouse canopy.

Many of the restorations will show the grinning-teeth paint job of the American volunteer Flying Tigers. The P-40, along with the P-39 Cobras, carried the Army Air Force through the first two years of WWII. Ruggedness and reliability were more outstanding than speed or maneuverability; many in Allied air forces; most saw action in the Pacific theater. The final and most numerous production model was the P-40N (illustrated).

Chance Vought F-4U Corsair

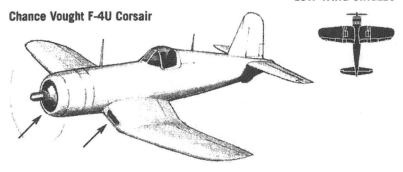

P-51B Mustang

North American P-51D Mustang

Curtiss P-4ON

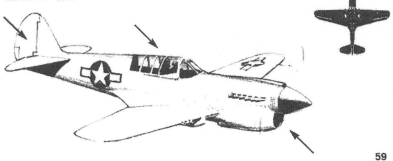

✈ HIGH-WING SINGLES

de Havilland (Canada) DHC3 Otter
Length: 41'10" (12.8 m) **Wingspan:** 58' (17.69 m) **Cruising speed:** 121 mph (195 km/h)

Fairly common in the Far West, Alaska, and Canada. **Massive single-braced high-wing tail dragger, with huge radial engine**; nearly two-thirds the size of a DC3. If you've never seen a de Havilland Beaver or Otter before, note the passenger windows: Otters show six rectangular side windows behind a cockpit window configuration that's similar to the much smaller Beaver's.

Built from 1952 to 1967, this late design carries the most massive, antique-appearing tail assembly of any post-WWII aircraft. Essentially an upscaled Beaver (the design project was called "King Beaver"), it carries up to 10 passengers. Single 600 hp radial engine proved quite reliable, even in the Arctic. Not uncommon on floats, particularly with small Alaskan and Canadian air-taxi operators.

de Havilland (Canada) DHC2 Beaver, U-6
Length: 30'4" (9.24 m) **Wingspan:** 48' (14.64 m)
Cruising speed: with radial, 135 mph (217 km/h); with turboprop, 157 mph (253 km/h)

A common float plane; less common elsewhere. Massive **single-braced high wing**; much more common with radial engine (upper drawing). Land versions with **fixed one-rung ladder**. Factory-standard **float planes with multirunged ladder and curved ventral finlet under tail fin. Trapezoidal passenger window with trailing porthole window** is typical on all models.

Built from 1948 to 1969; seats up to eight, including the pilot. All-metal construction. Numbers of them have crashed and been totally rebuilt. The less common turboprop (built between 1964 and 1969) also introduced the swept tail fin of modern design, as it did a fuselage lengthening that put the cockpit forward of the wing (lower sketch).

Noorduyn Norseman, C-64
Length: 32'4" (9.86 m) **Wingspan:** 51'8" (15.75 m) **Cruising speed:** 141 mph (227 km/h)

Extremely rare; seen most often in Canada. With its **huge single radial engine** and typically on floats or skis, it could be confused with a de Havilland, but note the odd **bent landing gear**, which always shows whether above wheels or floats or skis; **rounded wing and tail plane tips;** and **odd geometry and layout of windows. Deep fuselage** usually reveals its structure: **fabric over metal tubing.**

Built in various models from 1937 to 1950, with more than 700 in U.S. armed forces during WWII (USAF C-64, USN JA-1). It was a premier short-haul airliner just after the war and survives in limited numbers throughout the northern woods and lakes. Carried 10 in military discomfort (bucket seats) and 6 in upholstered airliner seating.

de Havilland DHC3 Otter

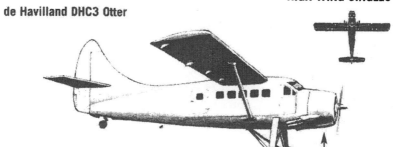

de Havilland DHC2 Beaver

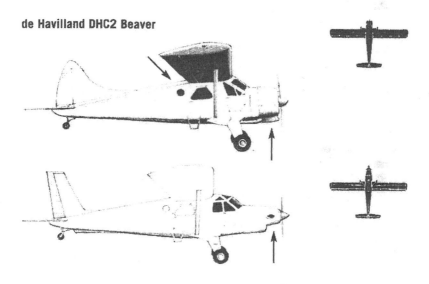

Noorduyn Norseman, C-64

Cessna 190/195 Businessliner
Length: 27'1" (8.26 m) **Wingspan:** 36'2" (11 m) **Cruising speed:** 160 mph (257 km/h)

Rare. A unique combination of **a tail dragger with skinny spring-steel wheel struts; big radial engine in a bumpy cowling; all-metal skin; and unbraced high wing.** Nothing else puts all that together.

A four-place luxury plane built from 1947 to 1954, the largest, fastest, roomiest, and easily the most expensive of the early postwar private planes. Model numbers refer to type of engine. A factory-standard float plane incorporates a three-finned tail, instead of the usual single tail fin, for lateral stability to overcome the wind drift on the floats—a tail like a miniature version of the Lockheed Constellation (page 154).

Howard DGA15, Nightingale
Length: 24'10" (7.57 m) **Wingspan:** 38' (11.58 m) **Cruising speed:** 180 mph (290 km/h)

Very rare. Everything about this plane is heavy, oversized: **large radial in smooth cowling; big propeller spinner; heavy gear, always with wheel pants; fixed two-rung ladder; tall tail fin;** nearby, the **V-struts enter a distinct underwing fairing.**

Developed from a long-distance racer design, the D(amn) G(ood) A(irplane) 15 was produced from 1939 (50 civilian versions) to 1942 (500 military models). Exceptionally roomy, it was a flying ambulance for the Navy (Nightingale) and a multipurpose trainer. High powered; not easy to fly; not particularly forgiving. Its printable nickname was Ensign Eliminator.

Curtiss–Wright Robin
Length: 24' (7.31 m) **Wingspan:** 41' (12.5 m) **Cruising speed:** 85 mph (137 km/h)

One of the rarest high-wing planes illustrated. **Enormous wing,** not only long but also with a 6 ft. constant chord. Curious **wing braces are parallel with several auxiliary struts. Big wheels** on the main gear; **squared-off trailing edge to tail fin** is unusual in such an antique aircraft.

Douglas "Wrong-Way" Corrigan, who had worked on Charles Lindbergh's *Spirit of St. Louis,* made the Curtiss–Wright Robin forever immortal (accounting for the large interest in restoring and recreating the 1928–1930 aircraft) by "accidentally" flying one from Long Island, New York, to Ireland in 1938; he always maintained that he was trying to fly nonstop to Los Angeles but that his compass reversed, and he flew 180 degrees off course. Built to seat three: the pilot, followed by a pair of wicker seats that could be offset to keep shoulders from rubbing. Corrigan flew his from a rear seat, peering over an auxiliary gas tank in the front seat.

Cessna 190/195 Businessliner

Howard DGA15

Curtiss–Wright Robin

Stinson Reliant, AT-19 (V77)

Length: 27'10" (8.48 m) **Wingspan:** 41'11" (12.77 m) **Cruising speed:** 120 mph (193 km/h)

Uncommon. **A massive, braced high-wing, always with cowled radial engine.** Typical wing has a single strut; earliest models, a pair of almost parallel struts. **Unique wing shape: swollen over strut area, giving the illusion of a gullwing.** Earliest models also have a "corrugated" cowling; typical surplus AT-19s; all late models have a smooth cowling.

Stinson Reliants appeared in 1935, continuing until 1942 as the Lend-Lease trainer and transport designated AT-19, used for radio and radar training in Great Britain. One of the earliest four- to five-seaters, it was not an uncommon short-haul airliner and company executive plane. A few battered models still flying as bush planes.

Monocoupe 90

Length: 20'6" (6.25 m) **Wingspan:** 32' (9.75 m) **Cruising speed:** 115 mph (185 km/h)

Quite rare. Something about this **V-braced, high-winged, radial-engined** aircraft catches the eye. It is **extremely short with a wide cabin and a very narrow rear fuselage; cowling bumps over cylinder heads; very small propeller spinner.**

Designed in Moline, Illinois, in the golden age of amateur enthusiasm; built from 1930 to 1942. Extremely agile little plane, used successfully in aerobatic and closed-course racing during the 1930s. Once the most popular high-performance small plane, it sat two in side-by-side comfort. Charles A. Lindbergh, who could fly anything he wanted, owned a Monocoupe.

Fairchild 24, UC-61 Forwarder (Argus)

Length: 23'9" (7.23 m) **Wingspan:** 36'4" (11.07 m) **Cruising speed:** 120 mph (193 km/h)

No longer common. **Roomy, high-backed fuselage** gives the impression of a small airliner; **V-braced high wing has a return strut to the wing root; notch (for visibility) in wing over windshield is unique,** as is the landing gear brace: **one wheel brace from fuselage, other from wing brace.**

Built from 1932 to 1947, including several hundred wartime UC-61s. About half the production was with a large radial engine, but most of those still flying are the illustrated in-line types. However, the field marks are consistent. Unusually roomy interiors sat four in military and post-1938 models. The sleek design was influenced by Raymond Loewy, creator of the Coke bottle and the Super Chief train.

Stinson Reliant, AT-19

Monocoupe 90

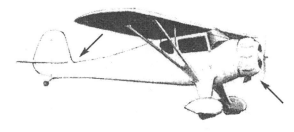

Fairchild 24, UC-61

Rearwin Skyranger

Length: 21'9" (6.6 m) **Wingspan:** 34' (10.36 m) **Cruising speed:** 100 mph (161 km/h)

Very rare. This small, fabric-covered, high-winged tail dragger is best singled out by a disproportionately large tail fin and single side window.

Never manufactured in large numbers (some 350 between 1940 and 1946), the little Skyranger was a comfortably furnished sport plane that came on the market when most manufacturers were dedicating their efforts to the pre-WWII pilot-training programs. Sat two, side by side, with standard dual controls; for the time, an unusual "slotted" wing gave aileron control at exceptionally low speeds. It has a landing speed of 48 mph (60 km/h).

Fleet Canuck

Length: 22'5" (6.83 m) **Wingspan:** 34' (10.36 m) **Cruising speed:** 85 mph (137 km/h)

Rare except in Canada. Not just another V-braced constant-chord high-wing. A much jauntier look than the similarly sized Piper Cub; more like the very similar Taylorcraft Model B (page 74). Close by, note the rectangular side window with trailing triangular quarter window. Before deciding, compare windows and tail fin shape with Taylorcraft.

Just over 200 built from 1946 to 1951. A popular light bush plane and a common club and trainer for Canadians: the least expensive plane available and built in Canada to boot. Somewhat overbuilt for strength, it was not certified for aerobatics, but more than one owner has looped it. Hard to stall or spin, with a leisurely landing speed of 44 mph (59 km/h).

Stinson Sentinel, L-5

Length: 24'1" (7.34 m) **Wingspan:** 34' (10.37 m) **Cruising speed:** 115 mph (185 km/h)

Rare. One of the few aircraft whose total impression is more distinct than individual field marks. The relatively massive, sweeping tail, much like a B-17 tail fin; the upturned nose; and the sweeping belly curve from nose to tail are distinctive. Close by, note the unique cross-bracing of the side windows, making three triangular panes. A very few of these planes have been converted by civilian owners to normal-looking cockpit canopies.

From 1941 to 1944, 5,000 were built. The Flying Jeep was the second most common "grasshopper" in the U.S. armed forces, right behind the Piper L-4. Sat two in tandem, but with a hinged rear canopy it served as a flying stretcher-bearer. General George Patton, among others, had an L-5 as a personal aircraft.

Rearwin Skyranger

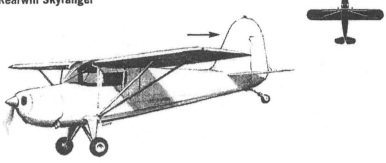

Fleet Canuck

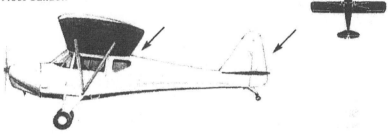

Stinson Sentinel, L-5

Cessna L-19 or O-1 Bird Dog, Ector Mountaineer
Length: 25'10" (7.89 m) **Wingspan:** 36' (10.9 m) **Cruising speed:** 105 mph (169 km/h)

Not common. An uncomplicated little single-brace, high-wing tail dragger; almost vertical windshield; wraparound rear window; curiously noncongruent side windows; noticeable (2.8-degree) wing dihedral.

More than 3,000 Bird Dogs were built from 1950 to 1958, many in civilian use. The Ector Mountaineer was a 1980s revival, built from off-the-shelf or reconditioned parts and more powerful engines. Ector also built in the float brackets as a standard item. Whether Bird Dog or Ector, the odd windows and the all-metal skin make it fairly easy to identify.

Maule Rocket M-4 to M-7
M-7 data: Length: 23'6" (7.16 m) **Wingspan:** 32'11" (10.03 m) **Cruising speed:** 164 mph (264 km/h)

A basic airplane produced in **dizzying variety:** aside from its general aspect, a **very long and very wide wing, built for a few years with a down-turned cambered wingtip** but not produced in that style today. Aside from the wing, its other obvious and unique field mark is the **V-brace to the wing, streamlined, with no return struts or wires.** The earliest M-4s did not have the wing camber. **Model 5s introduced the swept and rounded tail fin that is still the standard. Never seen with a radial engine. Recent cleaned-up cowling may show twin exhausts and a large air intake (turboprop) or single exhaust (piston engines).**

Essentially bench-built to order by the revived Maule company (Maule Air Inc.), you can have it your way: two or three side windows; fuel-injected, turbocharged, or turboprop engines; tail dragger or tricycle landing gear or factory floats. Interiors vary from Spartan to luxurious. The three-window Maules have huge doors for cargo loading. The most overengined and lighter early models could get airborne in 125 ft. (38 m); the later ones take up to 250 ft. (76 m), always with just a pilot and a half tank of gas. Designer Belford D. Maule once flew a Maule M-5 Super Rocket out of a hanger and was well airborne when he hit the doorway - -photo and details on the home page: www.mauleairinc.com.

Champion/Bellanca Citabria, Scout, Decathlon
Length: 22'8" (6.91 m) **Wingspan:** 33'5" (10.19 m) **Cruising speed:** 125 mph (201 km/h)

Of the small, **V-braced, constant-chord, square-end winged** planes, the Citabria is best distinguished by its **fancy wheel pants** and **squared-off tail fin.**

Champion Aircraft was manufacturing the tail-dragging Champion Traveller before it shifted to this version in 1964, with its more modern tail surfaces and wheel treatment, as well as strengthening, which made it certifiable as an aerobatic plane (Citabria is Airbatic, backward). One of the first planes capable of continuous inverted flight. From 1970 to 1980, Bellanca also built a nonaerobatic Scout and a strengthened, fully aerobatic Decathlon.

Cessna Bird Dog, L-19

Maule Rocket

Champion Citabria

Arctic Tern, Interstate Cadet (L-6)
Length: 24' (7.32 m) **Wingspan:** 36' (10.97 m) **Cruising speed:** 115 mph (185 km/h)

Not common. Another constant-chord, V-braced, high-wing tail dragger. A tandem-seat, slim plane whose most distinguishing feature is the tall, pointy tail fin, with noticeable trim-tab showing at tail plane level. New versions (upper drawing) have squared-off wingtips; older Interstates and L-6s have round tips. The 2-degree dihedral in the wing is, as usual, quite noticeable.

Very few of the originals survive, including the L-6 (not illustrated), which was an Interstate Cadet (lower sketch) with a greenhouse-type cockpit window. Interstate Cadets were produced from 1937 to 1942 as trainers; L-6s, until 1944. The design was revived in 1969 in Alaska, where the Arctic Tern (lower drawing) continues to be bench-built but with three visible changes: square wingtips, an angular rear-passenger window, and tail wheel moved all the way to the rear.

Funk (Akron) Model B to Model L
Length: 20' (6.1 m) **Wingspan:** 35' (10.7 m) **Cruising speed:** 100 mph (161 km/h)

Quite rare. One of the two braced high-wing singles with a pair of parallel braces (see the Porterfield Collegiate, page 72). Head on, the Funk engine cowling is unique, showing a round air intake completely surrounding the propeller spinner. Massive tail fin; squat, chunky overall appearance.

Built from 1939 to WWII and again from 1946 to 1948. A side-by-side two-seater that was considered remarkably easy to fly, responsive, and stable (note the large, high-lift wing and the substantial stabilizing tail assembly).

Stinson 10A (Voyager 90), Voyager 108, Voyager 108-1, -2, -3
Length: 22' (6.71 m) **Wingspan:** 34' (10.37 m) **Cruising speed:** 108 mph (174 km/h)

Not common and not just another braced high-wing tail dragger. Although the Voyager's general shape is unique, concentrate on some fairly trivial field marks for positive identification. All the Voyagers have a noticeable (2-degree) dihedral in the wing.

Voyager 90, model 10A (upper drawing): the two-seat side-by-side, with a possible third bench seat behind the pilot. The V-brace to the wing is quite unusual in that it has no supplementary cross or upbraces (contrast with a typical Piper Cub). Tail plane is set extremely low. Although distinctly a fabric-covered plane, the general effect is clean and neat, if stubby. Built from 1939 to 1942, when it was replaced by the military L-5 (page 66).

Voyager 108 (middle drawing): The four-seat Voyager, built from 1946 to 1948, looks much sleeker and slimmer than the Voyager 90 and has a longer engine cowling, housing an engine twice as powerful as the prewar Voyager's. Same simple V-brace without any supplements.

Voyager 108-3 (lower sketch): The last Voyager, with the much larger, vertical-style tail. Seats four. A few of the 108-3s were built by Piper until 1950.

70

Arctic Tern

Funk Model B

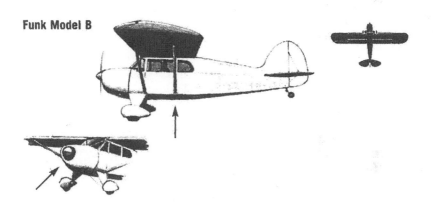

Stinson 10A

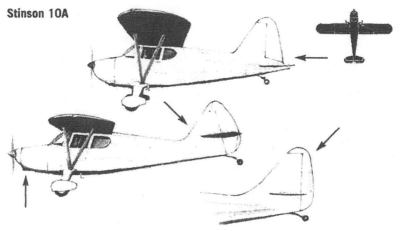

71

Porterfield Collegiate
Length: 22'8" (6.9 m) **Wingspan:** 34'9" (11 m) **Cruising speed:** 100 mph (161 km/h)

Quite rare. One of two **high-wing singles with parallel double struts.** Compare with Funk (Akron) Model B (page 70), a much chunkier, squatter aircraft with a larger tail fin. All fabric. If nothing was left of a Collegiate but the engine cowling, you could identify it by the **distinct cut-in for engine exhaust.**

A tandem-seat trainer and sportster; only about 500 built before WWII put Porterfield out of the airplane business and into manufacturing troop gliders in preparation for the invasion of Europe. As a trainer, extremely popular with students; with hands off, it would recover from spins or stalls and, for the nervous, could land at speeds as low as 40 mph (64 km/h).

Aeronca Champ, Traveller, Tri-Traveller, L-16
Length: 21'6" (6.56 m) **Wingspan:** 35' (10.66 m) **Cruising speed:** 90 mph (145 km/h)

Very similar to the Aeronca Tandem and the Aeronca Chief; separate from the Tandem by the Champ's **smooth engine cowling** and from the Chief by the slimmer fuselage/cabin, indicating its tandem seating.

Built from 1948 to 1964, the last dozen years by the Champion Aircraft Company, which acquired the design from Aeronca. Military observation versions (L-16) had four large, square side windows; otherwise, identical. Champion Aircraft called it the Traveller and also manufactured more than 1,000 Tri-Travellers, a popular flight-instruction model. The Tri-Traveller sits on its tricycle gear, with its nose distinctly turned up, quite noticeable on the flight line.

Aeronca Chief, Super Chief
Length: 21' (6.3 m) **Wingspan:** 36' (10.9 m) **Cruising speed:** 95 mph (153 km/h)

A pair of somewhat stubby, **braced high-wing two-seaters.** Like so many WWII planes, it's of **fabric construction,** with constant-chord wings and rounded tips. Close at hand, **Aeronca's trailing edge of the tail fin shows a noticeable extrusion—** an adjustable trim-tab. Once you've positively noted this, you'll find the shape of the entire plane sufficiently distinctive for long-range identification. The **Super Chief tail is much larger** (lower sketch). The Champion is very similar; its slimmer fuselage indicates the tandem seating for two. The original Chief was designed to take Continental's revolutionary opposed four-cylinder engine; first flown in 1938. With side-by-side seating for two, it was cozier than contemporary tandems, including the popular Piper Cubs. The Chief production ended in 1948. The Super Chief was built between 1946 and 1950.

Porterfield Collegiate

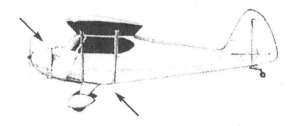

Aeronca Champ

Aeronca Chief

Super Chief

Aeronca 15AC Sedan
Length: 25'3" (7.7 m) **Wingspan:** 37'6" (11.43 m) **Cruising speed:** 114 mph (183 km/h)

A rare high-wing tail dragger: The single wing brace attaches much farther outboard than on somewhat similar Cessna high-wings. The distinctive tail fin appears to lean forward and shows the typical Aeronca bump.

Never common, the Sedan (close at hand, note the automobile-style door and window configuration) was built from 1947 to 1950 and came standard with dual controls. Perhaps 120 are still flying, some on floats. A roomy four-seater with good "high and hot" flying characteristics, it's capable of taking off with less than 500 ft. of ground roll at sea level.

Taylorcraft Model B, Taylorcraft F19, F21 Sportsman, F-22 Classic
Length: 22'1" (6.73 m) **Wingspan:** 36' (10.97 m) **Cruising speed:** 115 mph (185 km/h)

A variety of airplanes, based on a pre-WWII design, but in production as late as 1982. Large, upright tail fin with a distinct flat spot on the rudder; long, slim fuselage appears to "pinch down" to the tail assembly. Compare carefully with Taylorcraft Model D and L-2 Grasshopper (next entry). Lowest-priced Model Bs lacked the rear quarter-window. F-22 has tricycle gear.

The classic Model B Taylorcrafts (upper drawing), built from 1938 to 1958, lacked such niceties as wheel pants; so did the Taylorcraft F19 Sportsman, built by the revived company in 1968. Most sat two side by side, but a few were built in the 1950s to seat four. The revived Taylorcraft F19 and the wheel-panted (or, as they say in Britain, the "spatted-wheel") F21 (lower drawing) returned to the two-seater format.

Taylorcraft Model D, L-2, O-57
Length: 22'1" (6.73 m) **Wingspan:** 36' (10.97 m) **Cruising speed:** 90 mph (145 km/h)

Fairly common. Compare closely with the Taylorcraft Model B (previous entry), which too has the same large tail with a flat spot on the rudder. Always with **exposed cylinder heads** (but so were a few Model Bs). If tandem seating is visible, that separates it from the Model Bs; so does the A-shaped supplementary brace from the V-brace to the wing (Model B and F19 and F21 have a rectangular supplementary brace).

The L-2, with **greenhouse canopy and cut-down fuselage** (lower sketch), was a popular war-surplus purchase.

There was no advantage to retooling from the dual-control Model B trainers to the Model D Tandem trainer, except that it was the general wisdom that instructors should ride behind, not next to, the student. Several thousand Tandems and L-2s (also known as O-57s) were built from 1941 to 1945.

Aeronca 15AC Sedan

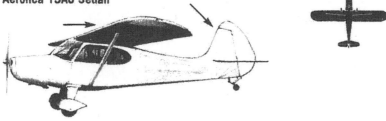

Taylorcraft Model B

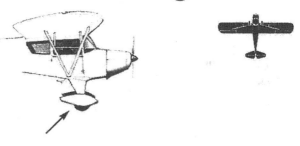

Taylorcraft F21

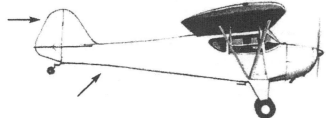

Taylorcraft Model D Tandem

L-2

Aeronca Tandem 65T, L-3

Length: 22'4" (6.8 m) **Wingspan:** 35' (10.6 m) **Cruising speed:** 80 mph (130 km/h)

Not common. Shares some field marks with early Piper Cubs. **Engine cylinders show through cowling** (as on Piper J3), but Tandem's cowling looks pug nosed. A small **triangular brace was added to the main wing braces. Tail rounded** (note flat spot on Piper J3 Cub tail). The rear-window shape is unique.

The Tandem was designed in 1940 for the pre-WWII Civilian Pilot Training Program— it's basically an Aeronca Chief with tandem seating. In a useful invention, the rear seat was suspended 6 in. higher than the front seat for visibility. The Army Air Force ordered thousands of Tandems with extra windows (lower sketch) as the L-3, a liaison and observation airplane.

Piper J3 Cub Trainer, PA11 Cub Special, J5 Cub Cruiser, PA12 Super Cruiser, J4 Cub Coupe, L-4

Length: 22'4" (6.8 m) **Wingspan:** 35'3" (10.74 m)
Cruising speeds: J3, 80 mph; Super Cruiser, 100 mph (129 km/h to 161 km/h)

Not every **constant-chord high-wing, fabric, tail dragger** is a Cub; it just seems that way.

J3 (upper drawing): **Exposed cylinder heads** (compare with Aeronca Tandem, L-3), **V-brace,** and **distinct flat spot on tail.** Some 5,000 built before WWII. A popular tandem-seat, two-person trainer that introduced nearly 75% of WWII aviators to flying, mostly through the Civilian Pilot Training Program. More than 5,000 built as L-4s for WWII observation-liaisons.

PA11 Cub Special, J5 Cub Cruiser, PA12 Super Cruiser (middle sketch): Despite a variety of engines and names, all are **three-seaters** (one pilot seat, with two passenger seats to the rear), with **fully enclosed engine.** Several hundred still flying, particularly the higher-powered Super Cruisers; many on floats. About 6,000 of the various three-seaters were built.

J4 Cub Coupe (lower drawing): Rarest of all. Compare closely with Super Cub (next entry) before deciding. **Engine cowling shows a distinct bump over cylinder heads** (compare middle sketch and Super Cub drawing), **a pudgy, dumpy look** caused by stuffing a side-by-side two-person cockpit onto the slim J3 Cub fuselage, which was designed for tandem seating. The J4 Cub Coupe tail is more rounded than that of the J3, making it quite similar to the Super Cub tail.

76

Aeronca Tandem

L-3

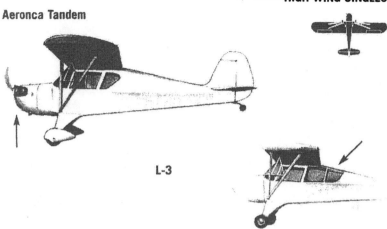

Piper J3 Cub Trainer

Cub Special

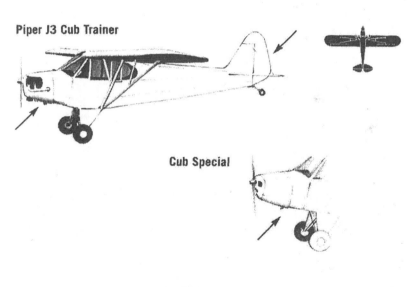

J4 Cub Coupe

Piper PA18 Super Cub, L-18
Length: 22'7" (6.88 m) **Wingspan:** 35'2" (10.73 m) **Cruising speed:** 115 mph (185 km/h)

Common as crabgrass. **Tail-dragging, all-fabric, rounded-tip, constant-chord, braced high-wing,** with **smooth cowling completely enclosing engine.** Compare with the J3 and Cub Cruiser (previous entry). **Always something showing below propeller spinner**—a location Piper has used for a variety of engine air intakes, landing lights, and so on, all absent on the earlier Cubs.

First flown in 1949 and kept in production (from inventory parts) as late as 1982, although dropped from Piper's official list that year. The success of the tandem two-seat Super Cub with standard dual controls was unquestioned—more than 30,000 were sold in the first 22 years of production. The Super Cub endured, but the various three- and four-seat Cubs were dropped in favor of new low-wing designs. The Super Cub, with more sophisticated construction methods (metal instead of wood wing spars, for example), is still essentially a power upgrade of the old tandem, two-seat J3.

AVIAT Christen Husky A-1
Length: 22'7" (6.88 m) **Wingspan:** 35'2" (10.73 m) **Cruising speed:** 140 mph (226 km/h)

Superficially, it looks as if someone dinged a Super Cub and replaced the wing with a Maule part. But note the **wide rectangular wing** combined with a more **old-fashioned-looking tail plane. Rectangular side windows.**

Christen took over the old Pitts factory (see both its sporting biplanes, page 10) in Wyoming and designed this new utility airplane from scratch. Now a division of AVIAT. The last cloth-over-tubing factory monoplane built. The U.S. Border Patrol replaced its Super Cubs with Huskies. Increasingly common on floats or with superwide tundra tires.

Luscombe 8A–8F, Silvaire
Length: 20' (6.09 m) **Wingspan:** 35' (10.66 m) **Cruising speed:** 105 mph (169 km/h)

Uncommon. A **small all-metal** plane, usually finished in **plain polished aluminum.** Strong men refer to it as "dainty" and "beautiful." Prewar models had fabric-covered wings. Wings show slight tapers toward the tip, separating it quickly from its constant-chord cohort. A distinct notch in the trailing edge of the wing over the cockpit is visible; it's similar to biplane upper wings. Compare with the Cessna 140 before being sure.

A pure sport and touring two-seater, designed in 1937 by Don Luscombe, author of the Monocoupe light-plane design. Only 1,200 built before WWII, but more than 5,000 built from 1945 to 1949 by Luscombe. A few more built by Temco; some bench-built by Silvaire as recently as 1960. Shown is the V-strut original 8A–8D models; 8E onward had a single strut.

Piper Super Cub

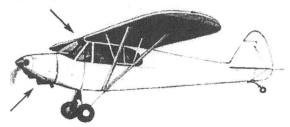

AVIAT Christen Husky A-1

Luscombe Silvaire

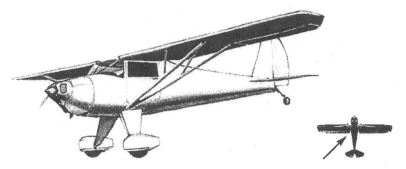

Cessna 120, 140
Length: 21'6" (6.58 m) **Wingspan:** 32'10" (10 m) **Cruising speed:** 105 mph (169 km/h)

Still common. A braced high-wing, tail-dragging single; most with two braces on a constant-chord wing with rounded tip. Deeply recurved tail planes, rounded tail fin. The model 120 was a stripped-down version, but the only visible difference is that the 120 lacks the quarter-window behind the passenger window. In 1949–1950, the 140D had the new all-metal Cessna wing and a single brace; it looks exactly like the model 170 (next entry) but with a smaller quarter-window behind the door and no dorsal fin fairing to the tail fin.

Introduced in 1946, the two-seat Cessna 120/140 was one of the least-expensive and highest-powered (85 hp) private airplanes you could buy. The spraddling spring-steel landing gear was so bouncy that the plane was actually more comfortable on grass strips than on paved runways, and it matched up nicely with the pasture pilots and small grass airports that were typical of the late 1940s. Nearly 5,000 built by 1950, when production ended.

Cessna 170
Length: 25' (7.62 m) **Wingspan:** 36' (10.96 m) **Cruising speed:** 110 mph (177 km/h)

Still common. An all-metal, tail-dragging, braced high-wing single with spring-steel landing gear. The rounded tail fin merging into a long dorsal fin is unique (other planes with the dorsal fin leading into the tail have more angular tail fins). A few (less than 10%) are early models with a constant-chord wing and two wing struts and without the dorsal fin: They resemble the 120/140 (previous entry) but are larger overall, with a much larger rear quarter-window.

The 170 was essentially a trade-up to four seats from the extremely popular Cessna 140. After 1 yr. (1948), the company introduced the all-metal tapered wing and subsequently sold nearly 5,000 170s. It became the Cessna 172 after 8 yrs. production by the simple addition of a tricycle gear and an angular, less romantic tail fin. Some 170s, meant for paved-only use, have wheel pants on the main gear.

Piper PA20 Pacer, PA22 Tri-Pacer, PA15 Vagabond
Length: 20'4" (6.2 m) **Wingspan:** 29'4" (8.9 m) **Cruising speed:** 130 mph (209 km/h)

A set of braced high-wing singles with two struts to wing (compare with similar Cessnas, with a single brace). Wings similar in shape but much stubbier than on the Piper Cub and Super Cub. The Tri-Pacer (upper drawing) also shows a large air scoop over the nose gear.

Piper, which had been building the very successful tandem-seat Cub series, decided to add another low-cost item in 1948 and 1949, the fabric-winged PA15 Vagabonds, side-by-side two-seaters. These quickly grew into the four-seat Pacers, with more powerful engines than the Cubs. The much stubbier Pacer wing (about three-quarters the total area of the Cub wing) did allow the Pacer to fly about 20 mph faster than the comparable Cub.

Cessna 140

Cessna 170

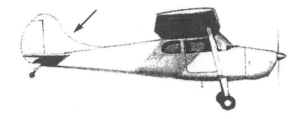

Piper Tri-Pacer

Piper Pacer

Cessna 180/185 Skywagon, Carryall, Agwagon
Length: 25'9" (7.85 m) **Wingspan:** 35'10" (10.92 m) **Cruising speed:** 129 mph (208 km/h)

A large tail dragger with braced high wing. Size and the presence of three side windows separate it from the 140 and 170 (page 80). Has a substantial tail—slightly smaller on the model 180 than on the 185—but this is difficult to determine the first time unless the planes are side by side. After you've seen them both, it's quite noticeable.

Produced for 30 yrs., 1953–1983, with minor changes (windows, engines, and making the drooping wingtip standard on recent models). The big-tailed, six-seat 185, first produced in 1961, is a very common float plane in the north woods. Standard spray-boom-equipped models for agricultural use show not only the booms but also a 160 gal. spray tank that attaches to the fuselage under the cockpit. The slight (less than 2-degree) dihedral in the wing is quite noticeable.

Helio Courier, U-10
Length: 31' (9.45 m) **Wingspan:** 39' (11.89 m) **Cruising speed:** 150 mph (241 km/h)

Not common. Unbraced high, constant-chord wing; usually a tail dragger; a very few with fixed tricycle gear. On tail draggers, the forward gear is on extremely long struts and is set well forward of the wing. Very tall, upright tail fin.

Manufactured from 1955 to 1978; about half the small production went to the U.S. Air Force as U-10s, a common liaison, cargo, and anti-insurgency plane in the Vietnam War. The only airplane completely designed by Harvard and Massachusetts Institute of Technology faculty members. Full-length leading-edge slotted flaps and massive slotted trailing-edge flaps give it a bizarre, short takeoff and landing capability. Seats up to six. Whatever the gear or engine type, the tail and wing configurations are consistent.

Pilatus PC-6 Turbo Porter, UV 20A Chiricahua
Length: 36'1" (11 m) **Wingspan:** 52'1" (15.87 m) **Cruising speed:** 132 mph (213 km/h)

The most rectangular plane in the air. Hershey bar wings and tail planes; fuselage cross-section is also rectangular; curious arched windows in passenger door; very long nose.

Modern, high-performance STOL (short takeoff and landing) aircraft. A few early Turbo Porters had a shorter nose than the PC-6, but they still don't look like a turboprop de Havilland Beaver (page 60). A very few with radials still flying. When with landing gear, the wheels are unusually large and on spidery, high struts. Movie fans may remember it as the bad guys' float plane in *Never Cry Wolf*.

Cessna 180/185 Skywagon

Helio Courier

Pilatus PC-6 Turbo Porter

Cessna 150, 152
Length: 24'1" (7.34 m) **Wingspan:** 33'2" (10.11 m) **Cruising speed:** 120 mph (193 km/h)

A series of **small, two-seater, braced, high-wing** planes; commonly fitted with dual controls for training. From 1970 onward, an optional version (the Aerobat) had structural strengthening for aerobatic flying; these will have a pair of cockpit-ceiling-through-the-wing windows. Some 30,000 150s and 152s were built (most of them resembling the upper drawing). Many converted to tail draggers.

Model 150A, B, C (lower drawing): Note the **two side windows** and the **upright tail fin.** About 3,000 were built from 1959 to 1963.

Model 150D (not illustrated): Built only in 1964; has the single side window and wraparound rear window of the late Model 150s and all Model 152s (upper drawing) but with the upright tail fin of the earlier 150s.

Model 150s built from 1965 to 1977, and all Model 152s built after 1977: show **single side window, wraparound rear window, swept tail fin.** The 1965 150Es had a shorter dorsal fin fairing into the swept tail.

OMF Aircraft Symphony 160
Length: 22'10" (6.96 m) **Wingspan:** 35' (10.67 m) **Cruising speed:** 145 mph (237 km/h)

A **very small aircraft;** of German design from a builder of motorized sailplanes, it shows its heritage in the **large wing for such a small plane** and the **proportionally huge tail fin and rudder** that rise out of a very slimmed-down fuselage. Quite **large windows** for such a small plane. Fixed landing gear (with wheel pants) cantilevers well out from the fuselage for a **wide stance.**

This small and economical plane has sold a few copies in U.S. flight schools, where it is one of the lowest-priced trainers on a per hour basis. The German parent company went bankrupt, and OMF, using government funds, built a manufacturing plant in Quebec province that shut down "temporarily" in 2003 after building half a dozen. Its future is particularly undecided.

North American Rockwell Darter Commander, Lark Commander
Length: Lark, 27'2" (8.28 m) **Wingspan:** 35' (10.67 m) **Cruising speed:** 130 mph (209 km/h)

Rare. **Constant-chord wings, with square tips, tricycle gear.** Darter Commander (upper sketch) is 5 ft. shorter and has **upright angular tail fin.** Lark Commander (main drawing) stretched the fuselage and added **swept tail fin.**

Odd little four-seaters: designed by the Volaire company, which was acquired by Aero Commander, which was acquired by Rockwell. From 1968 to 1971, Rockwell built fewer than 2,000, as the parent company switched to low-wing designs in single-engine aircraft (the Aero Commander 112). Intended to compete with the Cessna 150, of which more than 10,000 had been delivered before the Darter/Lark came on the market.

Cessna 152

Cessna 150

OMF Aircraft Symphony 160

North American Rockwell Darter Commander

North American Rockwell Lark Commander

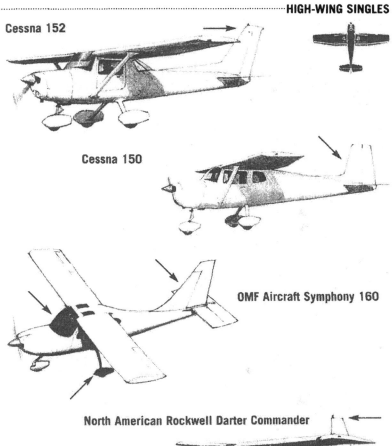

85

Cessna 172, 172 Skyhawk, T-41 Mescalero, 175 Skylark, Cutlass, Cutlass RG, Hawk XP

Length: 27'2" (8.28 m) **Wingspan:** 36'1" (11 m) **Cruising speed:** 172 Skyhawk, 140 mph (225 km/h)

Ubiquitous. A series of classic high-wing single-engine Cessnas produced for 30 yrs. We'll take them in order, from the 1956 introduction of the Cessna 172, essentially a 170 with tricycle landing gear.

Cessna 172 (upper drawing): **Two side windows; no rear window; high, unswept tail fin, with corrugated rudder. Squared-off nose** (compare with the 182/Skylane cowling, small sketch to the left).

Cessna 172 Skyhawk (model years 1960–1963) and 1958 model year Skylark: This is the old 172 cabin configuration with **swept tail fin** and **wheel pants.**

Cessna 175 Skylark (1959–1962): Until maintenance problems killed the idea, the Skylark was distinguished by a geared-down propeller. Note the **hump behind the propeller spinner;** otherwise identical to contemporary Skyhawks.

Cessna 172 Skyhawk (after 1964): Drawing shows the latest model, with a **long dorsal fin fairing to tail fin and a wraparound rear window.** The dorsal fin, shorter when the plane was introduced, reached this length in 1971. Distinguish this model from same-age 182 Skylanes, which have a flat rear window. Skylanes are also bulkier and huskier than Skyhawks, but you should make the distinction close at hand and then learn the conformation. Some 172s seen in blue and white paint, with "U.S. Air Force" lettered on the side but without other insignia, in civilian-operated contract flight schools near Air Force training bases, where it is the 30-hr. primary trainer, designated T-41 Mescalero.

Cessna Hawk XP (extra performance) (after 1978): A 172 Skyhawk with fixed gear, a more powerful engine, and subtle differences in only the nose cowling. Note the **larger spinner** and the **sleek cowling, with landing lights just above the nose wheel.**

Cessna 172 Cutlass: A 180 hp version of the 172 Skyhawk; no visible differences.

Cessna 172 Cutlass RG: **A retractable-gear Skyhawk; wheel wells remain open.** Distinguish from the very similar but bulkier retractable Skylane RG by the wraparound rear windshield. After you've seen them both close at hand, the difference in their shape will be a better field mark.

Cessna 182
Skylane (pre-1960)

Cessna 172 (pre-1960)

Cessna 175 Skylark
(1959–1962)

Cessna 172 Skyhawk

Cessna 172 Skyhawk (post-1964)

Cessna Hawk XP

Cessna 172 Cutlass RG

87

Cessna 182 Skylane, Skylane RG

Length: 28'2" (8.59 m) **Wingspan:** 35'10" (10.92 m) **Cruising speed:** 157 mph (253 km/h)

A pair of identical, braced, high-wing singles. One, the RG, has retractable gear (upper drawing), which increases the cruising speed to 179 mph (289 km/h). When the gear is retracted, note the open wheel wells on each side of the fuselage. Skylanes have a **flat, nonwraparound rear window.** Compare with the Cessna 172 series (page 86): Early 172s lacked rear window; later types have wraparound rear window.

What we have here is a more powerful version of the older model 172. But Cessna already had that in the model 180 (page 82).

Cessna Stationair, Skywagon, and Super Skylane

Stationair 7 data: Length: 31'9" (9.68 m) **Wingspan:** 35'10" (10.92 m)
Cruising speed: 156 mph (251 km/h)

Common, variable. A series of pilot plus five- or six-passenger aircraft most easily distinguished from their Cessna stablemates by sheer size; **all with single brace, wheel pants, and swept tail fins.**

Cessna 205 and 206 (Stationair 5, 6) and Super Skylane (upper drawing): **three passenger windows.** The more comfortably appointed Super Skylane looks just like a Stationair 6 from the port side but has a single door, not the double cargo doors of a Stationair, on the starboard side. This group has the same wing but a fuselage length of 28 ft. (8.53 m).

Cessna 207, 208 (Stationair 7, 8): Noticeably longer, emphasized by the **four, not three, side windows.**

Until the invention of the 14-passenger Caravan, the Stationair 8 was the largest braced-wing Cessna and one of the larger single-engine planes made. It will come as no comfort to those who try to put the proper names on things to learn that the original model 206 was called a Skywagon; that the next version, the model 207, was also given that name; and that the 206 was then called a Stationair again. When the final version of the 207 came out, it was called a Stationair 8. Several thousand of all types have been built since 1964. For the real Skywagon, see page 82.

Cessna Skylane RG

Cessna 182 Skylane

Cessna Stationair 6

Cessna Stationair 7, 8

Cessna 210 Centurion, Turbo Centurion
Length: 28'2" (8.59 m) **Wingspan:** 36'9" (11.2 m)
Cruising speed: Centurion, 193 mph (311 km/h); Turbo, 222 mph (357 km/h)

An unbraced high-wing. The tail plane is mounted slightly higher than on Cardinal series; two large side windows on Centurion, four small windows on pressurized Turbo Centurion (but compare with Cardinal RG, next drawing). Almost all Centurions have a dorsal fin that begins at the rear of the cabin (compare with shorter fin on Cardinal RG).

Seating the pilot plus six, the Centurions first flew in 1967. Their combination of unbraced high wing and retractable gear, along with the Cardinal RG, is unique in the industry. The pressurized Centurion was added to the line in 1977. A few early models, built from 1964 to 1966, have a braced wing and are virtually indistinguishable from a Cessna Cutlass RG (page 86). If you see an unbraced-wing Centurion that appears to have a smaller dorsal fin than illustrated (or happen to see a pair of them parked side by side), it is one of the models built in 1967 or 1968. Centurions built from 1969 to 1978 had doors to cover the main landing gear. Models built after 1979 eliminated the doors and show a distinct notch just under the rear of the cabin (typical as on lower sketch of the T210 pressurized Centurion).

Cessna 177, Cardinal, Cardinal RG
Length: 27'3" (8.31 m) **Wingspan:** 35'6" (10.82 m) **Cruising speed:** 177RG, 139 mph (224 km/h)

A small airplane with an unbraced high wing and retractable gear is either the 177 Cardinal RG or the slightly larger Cessna 210 Centurion (preceding entry). Both Cardinal models have a very low-mounted tail plane that, compared to the Centurion's inserted tail plane, looks as though it has been glued on. The Cardinal's fairing to the tail fin begins aft of the cabin; compare with the Centurion's fairing, which begins at the cabin.

Of the 4,290 of these dapper planes built from 1967 to 1978, more than 1,300 were the retractable model. As with all Cessnas, the vanilla version simply has a model number; names, such as Cardinal, indicate some enhancements, including in this case more horsepower, fancier interiors, and full blind-flying instrumentation. The unbraced wing looks nice but didn't add much speed, and Cessna was essentially competing with its own very popular braced-wing Cessna Skylane (page 88). The Cardinal was a success, a tribute to Cessna's skill at marketing very similar aircraft, something like the Big Three automakers.

Cessna 210 Centurion

Turbo Centurion

Cessna Cardinal RG

Cessna Cardinal Classic

Gippsland Aeronautics GA8 AIRVAN
Length: 29'4" (9 m) **Wingspan:** 40'8" (12.4 m) **Cruising speed:** 140 mph (224 km/h)

The only other high-winged planes with a fairing to the tail fin that **rises abruptly from the fuselage** are the 8'4" (2.54 m) longer Cessna Caravan (next entry) and the new 12'3" (3.73 m) longer Grand Caravan. **Strong dihedral in the wing** (none in the tail plane) looks as though the wing was folded in the middle and glued to the cockpit roof. **Single unbraced strut, slab sided, small undertail skid; odd rudder is short but deep.**

This Australian-designed (and built) carryall is meant for everything from hauling tourists to rough destinations (a cargo pack can be attached between the widespread wheels) to air ambulance and air surveillance with up to 9 hrs. slow-speed flight on station. The manufacturer saw a market opportunity when Cessna temporarily suspended production of the model 206 Stationair (page 88). The wing is the same basic design as used on its low-wing agplane, the GA200 Fatman, where it is mounted under the fuselage and up-braced.

Cessna 208 Caravan
Length: 37'7" (11.46 m) **Wingspan:** 51'8" (15.75 m) **Cruising speed:** 214 mph (344 km/h)

A **monster single**, comparable to the de Havilland Otter in size; **single Cessna-style brace to wing; five passenger windows; angular tail surfaces.**

The Caravan, with a single turbocharged 600 hp engine and carrying up to 14 people, is an attempt to find a replacement for the discontinued de Havilland Otters and Beavers and the many Cessna 180s and 185s. The tall fixed gear is meant for unimproved airstrips. Sales to military services are expected, as ambulance, parachute, and light transport. The Caravan can carry a ton and a half of freight more than 1,000 miles.

Cessna 208 B1 Grand Caravan, Super Cargomaster
Length: 41'7" (12.7 m) **Wingspan:** 52'1" (15.9 m) **Cruising speed:** 212 mph (341 km/h)

These largest Caravans are 4 ft. (1.22 m) longer than the decades-old and successful plain Caravans; at this size, the Grand Caravans equal the size of the original gigantic single, the long-discontinued radial-engined de Havilland Otter (page 60). These large Caravans have two more windows in the passenger models than the older Caravans and a large, optional cargo-belly bin on the blank-sided Super Cargomaster. (FedEx flies hundreds of Caravans and Grand Caravans with the belly bin.) The same single-braced high wing now shows, like all new Caravans, a radar domelet on the leading edge of the starboard wing.

To further distinguish the original Caravan from the new models, the old Cessna 208 is now, inexplicably, the Cessna 675. A few 208B2s are being built: Grand Caravans with plush seating for 7 instead of grimmer seating for 14.

Gippsland GA8 AIRVAN

Cessna 208 Caravan

Cessna 208 B1 Grand Caravan

Cessna Super Cargomaster

Extra Aircraft EA-500

Length: 32'6" (9.9 m) **Wingspan:** 37'9" (11.5 m) **Cruising speed:** 250 mph (402 km/h)

Very rare and trying to come into production as this book is published. Round and composite-smooth fuselage is bulky under the top-mounted wing; large tail fin with the horizontal stabilizer mounted very high; three passenger windows each side; strong skid strake.

A German aircraft that doesn't claim that it offers something extra; the designer is Walter Extra. As with many start-ups, refinancing and restructuring have delayed the introduction of the EA-500, which will sport a Rolls-Royce turboshaft driving a bulky five-bladed propeller. The round shape is not only easy to manufacture in plastics but also provides a strong pressure vessel with minimal engineering and manufacturing complexity.

Extra Aircraft EA-500

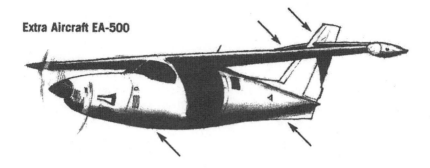

 **AMPHIBIANS**

Lake LA-4 Buccaneer, Renegade, Skimmer

Buccaneer data: Length: 24'11" (7.6 m) **Wingspan:** 38' (11.6 m) **Cruising speed:** 150 mph (241 km/h)

A series of four- to six-place amphibians. High-winged flying boat or amphibian with a **single engine mounted on a pylon high above the cabin; pusher propeller.**

Skimmer, the oldest version (center sketch), was a pure flying boat and lacked the support struts on the engine pylon. Buccaneer (upper drawing) is most common and is quickly distinguished from the Renegade by the lower position of the horizontal stabilizer/tail plane. The Renegade's tail fin control surface is all **below** the tail plane. Recent Lake Renegades (lower drawing) carry five passengers and are 4 ft. (1.22 m) longer than Buccaneers.

TSC1 Teal

Typical Teal II data: Length: 23'7" (7.19 m) **Wingspan:** 31'11" (9.73 m)
Cruising speed: 115 mph (185 km/h)

Very rare. Most flying are home-builts, but all are easily identifiable. **The only single-engined amphibian with a traction, pulling engine.** T-tail is also unique.

Original Teals came from the factory with standard dual controls, but most home-builts are single control with seating for three passengers. The original design had fold-up seats, on the presumption that the average user would be a fisherman who could turn the airplane into a john boat and fish right from the craft. Sold in kit form in the 1980s.

Republic RC3 Seabee

Length: 28' (8.53 m) **Wingspan:** 37'8" (11.48 m) **Cruising speed:** 105 mph (169 km/h)

A fat-cabined, thin-fuselaged amphibian with a **gently curved leading edge** to the tail fin. **Pusher propeller** mounted on the rear of the cabin.

On land, the Seabee is clearly designed as a tail dragger, and the rear wheel stays down in flight as the two front wheels retract up to, but not into, the fuselage. It was with visions of a vast postwar leisure-time market that Republic Aviation Company purchased Percy Spencer's design for his home-built Spencer S-12 in 1943 and certified the plane in 1946. It was an era when flying automobiles were being seriously designed as well. Republic cranked out 1,080 of the planes in a little more than 2 yrs., at a net loss of some $14 million. The mass market never caught up with the costs of tooling up and producing aircraft that sold for less than $6,000.

Lake LA-4 Buccaneer

Lake Skimmer

Lake Renegade

TSC1 Teal

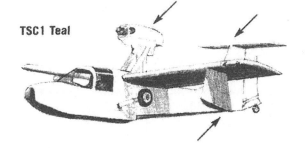

Republic RC3 Seabee

Grumman G21 Goose
Original data: **Length:** 38'4" (11.68 m) **Wingspan:** 49' (14.94 m) **Cruising speed:** 190 mph (306 km/h)

The oldest Grumman amphibian. **Fully rounded tail planes and fin** and twin engines that **angle out** slightly away from the centerline of the aircraft.

The Goose is such an old design (built from 1937 to 1946) that many owners have changed such details as cockpit and fuselage windows. Many fly today with turboprops replacing the old radials and with retractable floats that fold up and become part of the wing surface in flight. But the angled-out engine position remains despite all other modifications. Two crew and four to six passengers. Identifying the Goose is dependent on recognizing its Grumman origins and its old-fashioned boatlike lines. The somewhat similar Grumman Widgeon is noticeably smaller, and the very rare Grumman Mallard has a distinctively upswept look to the rear fuselage. See the next two entries before deciding you've seen the Goose.

Grumman G44 Widgeon
Length: 31'1" (9.47 m) **Wingspan:** 40' (12.19 m) **Cruising speed:** 130 mph (209 km/h) Mach 0.196

A small airplane with **in-line twin engines** mounted **parallel to aircraft midline;** sculpted Grumman-style fuselage.

Smallest of the twin-engine flying boats, the Widgeon saw extensive service as a patrol and antisubmarine craft in WWII. Although many have been converted to turboprops, the original Widgeon was sold with in-line engines, giving it a profile much different from the radial-engine Goose or Mallard. In most respects, the Widgeon is simply a scaled-down Goose, including the double-strut float mount; note, however, the less rounded tail fin and tail plane. Most of the 100 or so Widgeons still flying in North America have been converted by the McKinnon Company to turboprops and retractable wingtip floats.

Grumman G73 Mallard
Length: 48'4" (14.73 m) **Wingspan:** 66'8" (20.32 m) **Cruising speed:** 180 mph (290 km/h)

Rare. **Large,** with noticeable **upswept rear fuselage** and very **high tail fin;** large **radial engines and solid float pylons.**

Only 59 ten-passenger Mallards were built between 1946 and 1951. Look for one of the few remaining Mallards in Louisiana's bayou country and in the Bahamas. Most of these will have conversions to turboprop engines: Some have retractable floats. The only possible confusion is with the much larger (100 ft. wingspan) Grumman Albatross (next entry). The Albatross fuselage is massive compared to the Mallard's, and all Albatross noses show a distinct, protruding radar dome. As a luxury flying yacht, the Mallard flew for persons as diverse as Henry Ford and King Farouk of Egypt.

Grumman G21 Goose

Turboprop G44 Conversion

Grumman Widgeon

McKinnon T-Prop Conversion

Grumman G73 Mallard

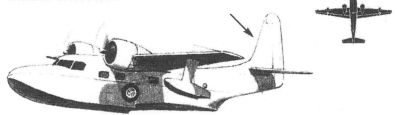

Grumman G64 Albatross

Length: 61'3" (18.67 m) **Wingspan:** 96'8" (29.46 m) **Cruising speed:** 225 mph (362 km/h)

Scarce. **Very large,** with **twin radial engines;** sculpted, curving fuselage; **cantilever wing** (no struts).

Another Grumman-looking aircraft, with solid pylons for the wingtip floats and huge radial engines. The Albatross was built for air-sea rescue, patrol, and antisubmarine warfare. Note the nose radar dome, which is not seen on the smaller Grummans. The Canadair CL215 (next entry) is almost as large as the Albatross, but compared to a Grumman design is all straight lines, whereas the Grummans have curves and shiplike moldings. Military versions were the HU-16 in the U.S. Coast Guard and the CSR-110 in the Canadian armed forces. Last military service was with U.S. Coast Guard; decommissioned in 1983.

Canadair CL215 and Bombardier Canadair 415

Length: 65' (19.8 m) **Wingspan:** 93'10" (28.6 m)
Cruising speed: CL215, 181 mph (291 km/h); Canadair 415, 234 mph (376 km/h)

This pair of water-scooping fire-bombers have several things in common: **huge size overall and very large wing length and wing area. Both have flat sides and angles instead of curves.** The original CL215 had radial engines and simple wings. However, some of those were **refitted with turboprop engines** and look similar to the new 415s. **To separate the all-turboprop 415s** (lower drawing) from the customer and factory refits, **note the small up-and-down tail finlets halfway from the huge tail fin to the outboard edge of the tail plane.**

A few Bombardier Canadair 415s are produced annually, making it the very last production (as opposed to specially fitted) amphibian in the world. Sturdy and maneuverable, it is successfully used in mountainous and windy terrain to get to forest fires before they get too large and go out of control. The turboprop version flies faster and farther than the original 215 and has the same carrying capacity for water: 10,800 lb. (4,899 kg), or 1,294 gal. (4,896 lit.).

Convair PBY-5 and PBY-6 Catalina

Length: 63'10" (19.5 m) **Wingspan:** 104' (31.69) **Cruising speed:** 130 mph (209 km/h)

Extremely rare. Huge **parasol wing** braced with **wing struts; twin radial engines.** The fuselage appears to hang suspended from the wing.

Although designed in 1935, the Catalina came equipped with retractable wing floats, something available only as postproduction modifications to Grumman flying boats. Most of the original PBYs were pure flying boats; most of the survivors are amphibious. Military PBYs had blister gun ports aft of the wings and a Plexiglas gun turret in the nose (or "bow"). The few civilian modifications still around have removed the forward gun turret, although a few kept the side blisters for sightseeing flights. The PBY-6, last of the series built, is identical to the PBY-5 except for a taller, thinner tail fin.

100

Grumman G64 Albatross

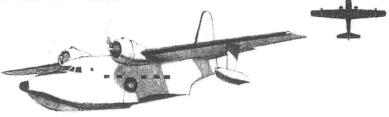

Canadair CL215

Bombardier Canadair 415 Super Scooper

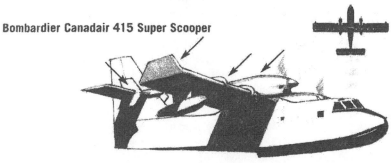

Convair PBY-5 Catalina

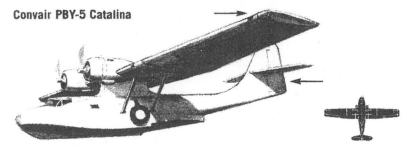

✈ LOW-WING TWINS

Diamond Aircraft DA42 Twin Star
Length: 27'10" (8.5 m) **Wingspan:** 44' (13.41 m) **Cruising speed:** 193 mph (311 km/h)

A very curious-looking, German-built, glider-influenced four-seater with **twin engines in deep, bulgy, nacelles; oversized one-piece windshield; fuselage goes into a wasp waist with strong dorsal and ventral fins** fairing into a **simple T-tail.** The abbreviated **winglets** are unique on a plane that flies so slowly.

This three-passenger (if good friends) twin is beginning to outsell the slightly smaller single-engined DA20s and DA40s. Like them, it offers engines for either JetA1 fuel (kerosene, basically) or regular diesel. This switch to economical fuel is no longer a selling point only in the European market. It's a flying roadster.

Beech Duchess 76
Length: 29'1" (8.86 m) **Wingspan:** 38' (11.58 m) **Cruising speed:** 175 mph (282 km/h)

Quite common. Small twin; **three side windows; one-piece curved windshield; Hershey bar T-tail plane and wing;** more **pointy nosed** than the comparable Piper Seminole; **distinct bullet on tail plane; engine nacelles stop at wing's trailing edge.**

Beech's entrant in the small four-seater twin market, used for multiengine training. First flown in 1974; first deliveries in 1977. The T-tail was extremely popular in the 1970s. Note the Piper Seminole and Cheyenne and the Beech Super King Air. The interest in T-tails was *perhaps* an affectation triggered by their wide use on jet airliners. Piper added T-tails even to existing single-engine models, the Lance and the Arrow. The Lance, however, reverted to a conventional tail, whereas the Arrow retained the T.

Piper PA44 Seminole, New Piper Seminole
Length: 27'6" (8.39 m) **Wingspan:** 38'7" (11.76 m) **Cruising speed:** 186 mph (300 km/h)

A small, common twin. **T-tail; flattened engine nacelles extend slightly behind wing; two-piece windshield; three side windows of irregular geometry** (compare with the small T-tail Beech Duchess 76 [preceding entry] before deciding). The other two T-tail twins are much larger; see the Piper Cheyenne III (page 122) and Beech Super King Air (page 120).

The Seminole (unrelated to the U.S. Army Seminole, its nickname for the military version of the Beech Queen Air) is a four-seat light transport and is popular as an inexpensive multiengine trainer. Came in a turbocharged version that is identical on the exterior but has an altitude ceiling of 20,000 ft. and a pressurized cabin.

Diamond DA42 Twin Star

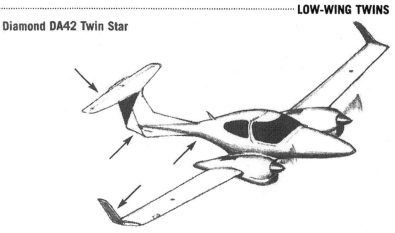

Beech Duchess 76

Piper PA44 Seminole

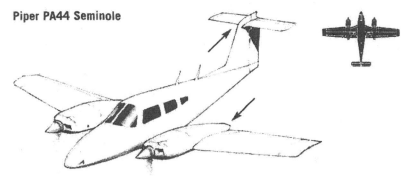

Piper PA23 Apache

Length: 27'3" (8.3 m) **Wingspan:** 37' (11.28 m) **Cruising speed:** 150 mph (241 km/h)

Increasingly uncommon. **An old-fashioned small twin, rounded tail fin, tail planes and wingtips; two (rarely three) side windows; small engines set close to fuselage;** retracted wheels stay slightly exposed and are visible.

Built from 1954 to 1960, the first really light twin with economical engines; seats four. The wheels, which do not quite retract, are built so deliberately (as on many WWII bombers). You can still land the plane if the system fails to extend the wheels; what's more, you can land, even if you forget to drop the wheels, without automatically demolishing the aircraft. Most restored models have higher-horsepower engines and slightly higher cruising speeds. A few models were built with three side windows.

Piper PA23 Aztec, PA23-235 Apache

Length: 31'3" (9.52 m) **Wingspan:** 37'3" (11.35 m) **Cruising speed:** 204 mph (328 km/h)

A family of similar aircraft. **Conventional tail, low-wing twin; swept angular tail fin; three side windows;** noses vary in length from short (PA23-235 Apache) to medium (Aztec B, C) (first two drawings) to long (Aztec D and later models; see silhouette). Seen overhead, the wing has complicated geometry: a Hershey bar shape but with added rounded wingtips and fairings from the fuselage to the leading edge of the wing at the engine nacelle and from the outboard side of the engine nacelle into the wing's leading edge. The last model, the Aztec F (lower drawing) has an angular outline to the wingtips, as though one had simply taken the old rounded shape and snipped it two or three times with a pair of shears.

Successor to the Apache (the first Aztec in 1960 was an Apache with a widened cabin to seat five and a new, angular, swept tail fin), the Aztec was a six-passenger twin available with turbocharged engines. An odd characteristic, occasionally useful as a field mark when the plane is overhead and going away, is that the tail fin and tail planes trail well behind the fuselage proper.

Piper PA23 Apache

Piper Aztec B

Piper Aztec C

Piper Aztec D

Piper Aztec F

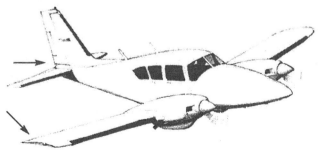

Grumman American/Gulfstream American GA7, Cougar

Length: 29'10" (9.09 m) **Wingspan:** 36'10" (11.23 m) **Cruising speed:** 190 mph (306 km/h)

Not common; look for it at airports offering multiengine flight school. Dihedral in wing and tail, combined with constant-chord wing; three side windows; swept tail fin.

First delivered in 1978, intended as an economical dual-control twin-engine trainer. Delivered as the Cougar with fancier interior. Seats four, including pilot and copilot or student. Production was sporadic, following the acquisition of Grumman American by American Jet Industries.

Piper PA34 Seneca I, II, III, IV, and New Piper Seneca V

Length: 28'7" (8.69 m) **Wingspan:** 38'11" (11.85 m) **Cruising speed:** 202 mph (326 km/h)

Common: Small low-wing twin, equal-chord Hershey bar wing and tail plane, swept tail fin. The whole tail assembly seems to be set behind the fuselage proper, quite visible overhead. Senecas II through IV had four side windows, each a distinctly different size and shape. (Seneca I had three larger side windows, also all different in size and shape.) Seneca I and II had a windshield with a center post; Seneca III (main illustration) got a one-piece windscreen; the last, Seneca V (lower sketch), has a somewhat flatter windshield. The largest difference between III–IV and V is the rationalizing of the side windows, which now line up top and bottom. Seneca Vs have the new Piper round air scoops in the nacelles.

Note the similarity in size and shape to the Piper Cherokee SIX (page 38). The Seneca's first trial flight as a twin was on a Cherokee SIX airframe that still had its single engine. It was flown as a trimotor, making the test aircraft the last nose and wing trimotor ever built.

Piper PA60 Aerostar, Ted Smith Aerostar

Length: 34'10" (10.62 m) **Wingspan:** 36'8" (11.18 m) **Cruising speed:** 231 mph (372 km/h)

Not common; unique design. A midwing twin; slight dihedral in wing, none in tail; leading edge of wing at right angle to fuselage, trailing edge tapers sharply to tip; tail plane strongly swept, bulbous nosed; wraparound windshield, with two small windows above cockpit; fairing to tail fin is cut off abruptly.

Ted Smith, a California designer, tried to build Aerostars from 1967 to 1978 in competition with the big three American builders. Although it's an attractive design and simple to construct, his company, after several reorganizations, ended up as the Santa Maria Division of Piper. Typical of the Ted Smith touch, the three swept tail surfaces (fin and planes) are interchangeable.

Gulfstream Cougar

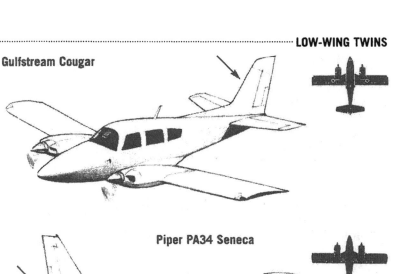

Piper PA34 Seneca

New Piper Seneca V

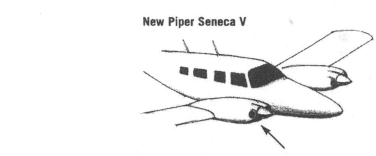

Piper PA60 Aerostar

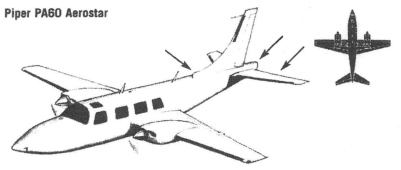

Beech 50 Twin Bonanza, L-23 Seminole
Length: 31'6" (9.6 m) **Wingspan:** 45'3" (13.8 m) **Cruising speed:** 203 mph (327 km/h)

A series of small, low-wing twins. Old-fashioned-looking vertical tail fin; dihedral in wing and tail; bulky engine nacelles house landing gear that does not retract fully. From two to four side windows, including the pilot's. But close at hand, note the unique three-piece windshield, with double divider strip in center.

Almost 1,000 of these stubby little aircraft were produced from 1952 to 1961. It was the first civilian twin-engine plane available after WWII and opened up the corporate airplane market. Engine horsepower varied from 260 to 340. Could hold six passengers in seats three abreast in its chubby cockpit.

Beech 95 Travel Air
Length: 25'11" (7.9 m) **Wingspan:** 37'10" (11.53 m) **Cruising speed:** 195 mph (314 km/h)

Fairly common. Very small low-wing twin; vertical tail fin; bulky nacelles; dihedral in wing, none in tail. Landing gear retracts completely; compare with Beech Twin Bonanza (previous entry). One-piece windshield. Close at hand, the triangular rear passenger window is unique, quite different from any Twin Bonanza.

Nearly 1,000 of these little twins, the lowest priced on the market, were built from 1958 to 1968. The plane had a single-engine service ceiling of 4,400 ft. above sea level, which effectively eliminated it from the substantial airplane market of the Rocky Mountain and intermountain West, where airports are typically above 5,000 feet.

Beech Baron D55, 58
Length: model 55, 28' (8.53 m); model 58, 29'10" (9.09 m) **Wingspan:** both, 37'10" (11.53 m) **Cruising speed:** both, 216 mph (348 km/h)

Common. A complex series of small low-wing piston twins. The consistent identification marks are the typical Beech wings, with a fairing from the wing root to the engine nacelle and dihedral in wing, none in tail. The 55 series had three side windows; the 58, four. The windshield is set forward of the wing's leading edge on the model 58, sometimes a useful field mark when the wing obscures the windows. A model 58 with turboprop engines, a swept tail plane, and a taller tail fin is the rare French-built Beech Marquis, a migrant from Europe.

A small four-place (three passengers and pilot) aircraft of considerable popularity. More than 6,000 delivered after 1960, including a few hundred of the stretched model 58 after 1970. Regular improvements were in engines, air conditioning, and avionics rather than in airframes.

Beech 50 Twin Bonanza

Beech 95 Travel Air

Beech Baron D55

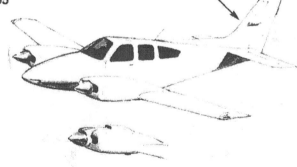

Beech Baron 58

Cessna T303 Crusader
Length: 30'5" (9.27 m) **Wingspan:** 39' (11.9 m) **Cruising speed:** 207 mph (333 km/h)

A low-wing twin, with the tail plane mounted well up the fin; long engine nacelles trail behind wing; three rectangular passenger windows each side; dihedral in wing, none in tail. Overhead, the wings and tail plane show symmetrical taper, with just a hint of the standard Cessna treatment; fairing from fuselage to wing's leading edge and from outboard side of engine nacelle to wing but much less visible than on older Cessna twins.

Cessna's 1982 entry into the fuel-economic, easy-to-maintain, piston-engine business-twin market. Long nose and trailing engine nacelles designed for baggage carrying. If you see it on the flight line, note that it's one of the few small twins with a drop-down stair built into the passenger door.

Beech B60 Duke
Length: 33'10" (10.31 m) **Wingspan:** 39'3" (11.96 m) **Cruising speed:** 250 mph (402 km/h)

A low-wing twin piston that shows strong dihedral in wing and tail; long, pointy nose; very strongly swept tail fin and tail plane; three rectangular windows each side. Does not have the trailing oval passenger window typical of so many Beech aircraft; compare with the Queen Air, King Air (page 120).

A four- or six-passenger plane with a crew of two but was frequently sold as a top-of-the-line personal aircraft and seldom used in the passenger business. Since 1968, delivered as a personal and corporate aircraft. It is easily recognized at a distance by its unique lines—the illusion of speed and a certain rakishness.

Rockwell (Fuji) Commander 700
Length: 39'5" (12 m) **Wingspan:** 42'5" (12.93 m) **Cruising speed:** 252 mph (405 km/h)

A low-winged twin; unswept and level tail plane mounted partway up fin; slim wings with dihedral; opposed-cylinder engines carried in flattened nacelles well forward of the wing; air scoops under nacelles for turbochargers. Trapezoidal passenger windows (three port, four starboard) are absolutely unique.

A joint design of Fuji in Japan and Rockwell International in the United States, it was first flown in 1975. Seats four to six in pressurized cabin and has a crew of two. Its practical range is more than 800 mi. (1,300 km). One of the few light twins built that used NACA (National Advisory Committee on Aeronautics) wing designs, although the slim and symmetrically tapering wings were constructed entirely in Japan.

Cessna T303 Crusader

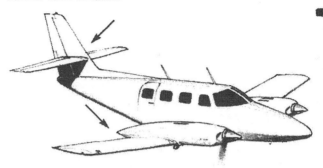

Beech B60 Duke

Rockwell Commander 700

Piper PA31P-350 Mojave
Length: 34'6" (10.35 m) **Wingspan:** 44'6" (13.35 m) **Cruising speed:** 270 mph (434 km/h)

A 1983 introduction. A low-wing twin, with turbocharged engines in very flattened nacelles that extend well behind the wing; dihedral in wing, none in tail; symmetrical tapering of both edges of wing and tail plane; three windows starboard, two port.

A five-passenger luxury business plane with piston engines seems an odd introduction in the turboprop era, but the intent was fuel economy and a power plant that could be worked on without a doctorate in engineering. The cabin is unusually deep for a small twin and is reflected in the bulky fuselage carried well aft. The long nose is for baggage, as are the trailing engine nacelles.

PIPER PA31 NAVAJO, PA31-350 CHIEFTAIN

A family of low-wing twins, with flattened engine nacelles housing opposed 6-cylinder engines; no tip tanks; all have characteristic Piper wing: distinct leading-edge fairing from fuselage to engine nacelle, both edges tapering from nacelle to wingtip; dihedral in wing, none in tail plane. Note that the newest Chieftain commuter (PA31-350 T1040) has turboprops in round nacelles that do not extend behind the wing--essentially like Cheyenne engines. Engine nacelles on the PA31-325 Navajo and the PA31-350 Chieftain extend beyond trailing edge of wing. Nacelles on PA31 Navajo and the pressurized PA31P stop well short of the trailing edge. Navajos carry 6 passengers; Chieftains can accommodate up to 10.

Piper PA31-325 Navajo CR
Length: 32'7" (9.93 m) **Wingspan:** 40'8" (12.4 m) **Cruising speed:** 244 mph (393 km/h)

Three large windows and one small side window, not counting pilot's side window; counterrotating propellers; nacelles extend beyond trailing edge.

Piper PA31 and PA31P

PA31 is identical to PA31-325 except that engine nacelles do not extend past trailing edge. PA31P (pressurized) has three windows starboard, two port (door on port side has no window).

Piper PA31-350 Chieftain
Length: 34'7" (10.55 m) **Wingspan:** 40'8" (12.4 m) **Cruising speed:** 251 mph (404 km/h)

The stretched Navajo is common in feeder airline and air-taxi service. Shows five windows on each side, not counting pilot's window; nacelles on most models extend beyond trailing edge. The less common PA31-350-T1040 has turboprops in round nacelles that do not extend past trailing edge.

Piper PA31P-350 Mojave

Piper PA31-325 Navajo CR

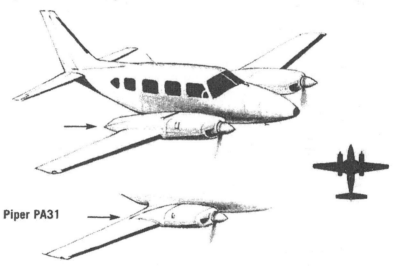

Piper PA31

Piper PA31-350 Chieftain

Piper PA31T Cheyenne
Length: Cheyenne IIXL, 36'8" (11.18 m) **Wingspan:** all models, 42'8" (13.01 m) **Cruising speed:** 244 mph (393 km/h)

Fairly common. Low-wing turboprop twin; engine nacelles blend into wing's trailing edge; swept tail fin; barely visible dihedral in wing, none in tail; tip tanks. The XL model illustrated has four passenger windows starboard, three port. Earlier models Cheyenne I and II are 2 ft. shorter and show three- and two-passenger windows, starboard and port. A few Cheyenne Is do not have tip tanks.

Built after 1969, the high-powered Cheyenne II was the typical and original Cheyenne. The Cheyenne I, a version with less powerful engines and less standard equipment, was not introduced until 1978.

Piper PA30, PA39, Twin Comanche
Length: 25'2" (7.67 m) **Wingspan:** 36'9" (11.22 m) **Cruising speed:** 186 mph (299 km/h)

A small low-wing twin. Manufactured with and without tip tanks; engine nacelles stop well short of trailing edge. Although it has the characteristic Piper fairing from fuselage to engine nacelles, the leading edge is straight and the trailing edge tapered, which gives the wing the illusion of leaning forward. Dihedral in wing, none in tail plane. Comes with two or (more commonly) three side windows, including the pilot's side window.

A successful and popular series that first flew in 1962. All seat four persons, including the pilot. Various models with turbocharged engines, counter-rotating propellers, and internal layouts. Models with tip tanks somewhat resemble the Cessna 310, but 310 nacelles extend beyond trailing edge, 310 wing has no fairing between fuselage and nacelles, and 310 shows two windows on each side, including the pilot's.

Cessna 310, 320 Skyknight, U-3, L-27
Length: 29'7" (9.02 m) **Wingspan:** 37'6" (11.43 m) **Cruising speed:** variable, about 177 mph (285 km/h)

A variety of popular aircraft sharing the minimum characteristics of twin engines on dihedral wing combined with level tail planes; very flat engine nacelles; tip tanks; distinct point at the bottom of the tail fin. Since 1969, there has also been a noticeable ventral fin (tail skid). Rare Skyknight has four small side windows. Close at hand, Cessna 310 and 320 tip tanks are distinctly canted up and out from the wing.

Cessna's entry into the business-twin market quickly became a military utility and liaison aircraft (U-3, L-27) and was produced continuously from 1954 to 1982. Model changes tended to emphasize minor changes in windows, streamlining, and engines. The 310s with ventral fin and without rear windows date from 1969 to 1973. The major change came in 1975, when the nose was lengthened and a turbocharged engine became available. The turbo versions cruise at more than 200 mph (322 km/h) and can be distinguished from the conventional engines by the absence of a cowl flap on the bottom of the nacelles.

Piper Cheyenne XL

Piper Twin Comanche

Cessna 310, 1969 model

Cessna 310, 1973 model

Cessna 310 Turbo

Cessna 340, 335
Length: 43'4" (10.46 m) **Wingspan:** 38'1" (11.62 m) **Cruising speed:** 212 mph (341 km/h)

A low-wing twin. Four small oval windows each side; noticeable ventral fin: long fuselaged and short nosed in its general aspect; engine nacelles extend past trailing edge, tip tanks are canted outward at a 30-degree angle; dihedral in wing, none in tail plane. Overhead, it could be confused with the smaller Cessna 310. These two Cessnas have straight leading edges on wings that arise directly from the fuselage, without any fairing there or at the engine nacelles, and have tip tanks.

This four-passenger, two-crew, pressurized aircraft has flown since 1971. The model 335 is not pressurized but has exactly the same window layout, giving no external evidence of its not being able to operate at 30,000 ft., as the 340 can.

Cessna 411, 414, and 421A, 421B Golden Eagle
421A data: Length: 33'9" (10.29 m) **Wingspan:** 39'11" (12.17 m)
Cruising speed: 226 mph (364 km/h)

A series of similar twins. Four or five passenger windows, **tip tanks, long noses, no ventral fin, strong dorsal fin fairing to highly swept tail fin.** All have the typical Cessna wing: straight leading edge, slight taper of trailing edge beginning at engine nacelles. **Dihedral in wing, none in tail plane.** Detail below main drawing shows the unpressurized model, the 411; note the single side window for the pilot (pressurized models have a two-part side window).

Beginning in 1965, with the unpressurized Cessna 411, then in 1967, with pressurized versions, a series of six- to eight-passenger twins was built until 1985. The 414 is a less expensive, lower-powered version of the 421. Models built from 1965 to 1972 show four round windows. From 1973 to 1985, the 421 had five oval passenger windows; the 414 added the fifth window in 1974.

Cessna 414A Chancellor and 421C Golden Eagle
Chancellor data: Length: 36'4" (11.04 m) **Wingspan:** 44'1" (13.44 m); Golden Eagle, 41' (12.5 m)
Cruising speed: 211 mph (339 km/h)

A pair of similar turbocharged twin-piston planes. **Five oval windows; dihedral in wing, none in tail plane; without tip tanks.** Very similar 414 Chancellor and 421A and the 421B Golden Eagle are identical, except with tip tanks. Compare with the almost identical Cessna Corsair, Conquest I (page 118), which has everything as in 414A and 421C except for a very sharp dihedral in tail and turboprop engines. That one company should make so many very similar models is curious and an annoyance to the viewer.

Cessna created two new models by dropping the characteristic tip tanks from its Golden Eagle and Chancellor series in 1976 (while continuing to manufacture planes with tip tanks). The new models, designated 414A Chancellor and 421C Golden Eagle, offered slightly better performance and some greater ease in managing the fuel systems.

116

Cessna 340

Cessna 421A

Cessna 421C

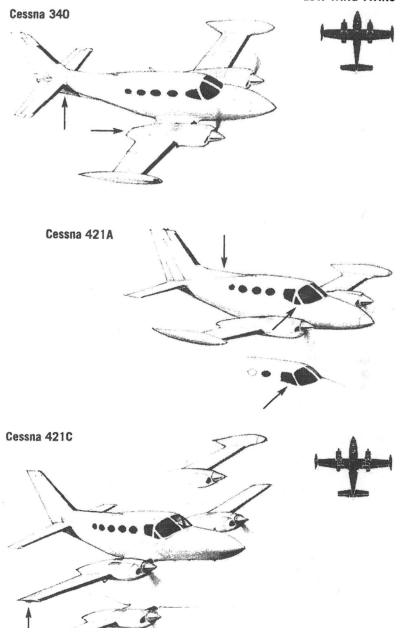

Cessna 401, 402 Utiliner, Businessliner
Length: 36'1" (11 m) **Wingspan:** 39'10" (12.15 m) **Cruising speed:** 200 mph (322 km/h)

A low-wing twin with that Cessna look: straight leading edge to wing; no fairing in wing at all; slight dihedral in wing, none in tail. Models built from 1967 to 1971 (401, 402A, and early 402Bs) have four evenly spaced round windows that get smaller toward the tail. Models from 1971 to 1987 (later 402Bs and 402C) have five rectangular windows on each side, also tapering in size front to rear. All 402Bs have tip tanks (see sketch).

Carrying a crew of one or two and six to nine passengers, Cessna Utiliners and Businessliners serve feeder lines and corporations. They aren't pressurized or particularly fast, but they were intended to be economical rather than exotic, as their sobriquets indicate.

Cessna 404 Titan, 406 Caravan II
Titan data: Length: 39'6" (12.04 m) **Wingspan:** 46'8" (14.23 m) **Cruising speed:** 230 mph (370 km/h)

The original Titan shows a very strong 12-degree dihedral in tail, separating it from the Cessna 401/402 (preceding entry), as does the number of passenger windows. Somewhat resembles the Conquests (next entry), but they have TV-screen (Conquest II) or oval porthole (Conquest I) windows, and they are both turboprops.

The last version of the Titan is the Cessna 406 Caravan II, which has an entirely different tail (see sketch): a low cross with no dihedral. Seen from the side, the Caravan II also has a distinct belly strake. In performance and efficiency, the Titan falls between the Businessliners and the Conquests and is scarcer than either of them.

Cessna 441 Conquest (Conquest II)
and Cessna 425 Corsair (Conquest I)
441 Conquest data: Length: 39' (11.89 m) **Wingspan:** 49' (14.94 m)
Cruising speed: 290 mph (467 km/h)

Both aircraft are low-wing twin turboprops. Very strong (12-degree) dihedral in tail plane. Except for the engine and the dihedral in the tail, the 425 Corsair (Conquest I) is identical to the Cessna 421C Golden Eagle. Corsairs are scarcer than DC3s. Overhead, a typical Cessna wing, unfaired at wing root or nacelles. Turboprop engines on the much more common 441 Conquest (Conquest II) extend far forward of the straight leading edge and do not show past trailing edge. Corsair (Conquest I) is similar, but it shows nacelle behind. Except when the plane is directly overhead, the dihedral will be very noticeable.

Cessna 402

Cessna 404 Titan, 406 Caravan II

Cessna 441 Conquest II

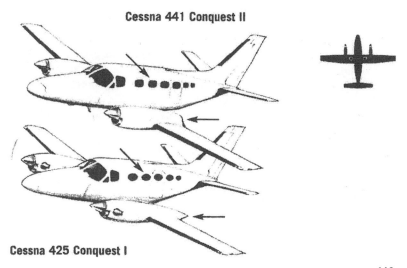

Cessna 425 Conquest I

Beech Queen Air, U-8, U-21 Seminole

Length: 35'6" (10.82 m) **Wingspan:** 45'10" (13.98 m) **Cruising speed:** 230 mph (370 km/h)

A series of midsized low-wing twins. Matching 7-degree dihedrals in wing and tail; strongly swept tail fin; three and four rectangular windows, port and starboard, with trailing small oval window. Earliest models (B65) had vertical tail fin.

Beginning with the Queen Air 65 in 1958, a long series of successful small twins with various engines. The matching dihedral is typical of both the Queen Air and the conventional-tail King Air and is an unusual combination.

Beech King Air A90–E90, A100, B100

Model E90 (includes U.S. Army U-21) data: Length: 35'6" (10.32 m)
Wingspan: 50'3" (15.32 m) **Cruising speed:** 260 mph (418 km/h)

A series of low-winged, twin turboprops with conventional tail. Slight dihedral in wings and tail plane. Typical Beech window details: no window in passenger door, one smaller window bringing up the rear, after a blank spot. Stretched A100 is 4 ft. longer than other models and has six large windows and one small window starboard; five large and one small, portside. Other models show four large windows and one small one on starboard; three large and one small on port.

More than 1,000 King Airs in service. The stretched A100 is a common feeder-line 12-passenger plane; the other versions, 6-passenger. Early King Airs were essentially pressurized Queen Airs with turboprop engines; easily distinguished overhead by the engine noise; on the ground, by the round pressurized windows fitted in the same pattern as the Queen Air's square passenger windows.

Beech Super King Air B200, 350, 1300, T-44, U-12

B200 data: Length: 43'9" (13.16 m) **Wingspan:** 54'6" (16.6 m) **Cruising speed:** 320 mph (515 km/h)

A low-winged twin turboprop with a T-tail. Compare with Piper Cheyenne III (next entry), and note that King Air has **round passenger windows, with the last one always smaller.** Don't rely on the standard Cheyenne tip tanks; many Super King Airs have optional tip tanks.

Earliest models showed four large windows on each side; later models, five to seven windows and the trailing smaller window. The newer B350 has winglets, extra windows, and is 34 in. (0.85 m) longer. Rare B1300s have a cargo-belly pod and ventral fins below the tail assembly. Common military VIP transport and twin-engine trainer.

Beech Queen Air

Beech King Air

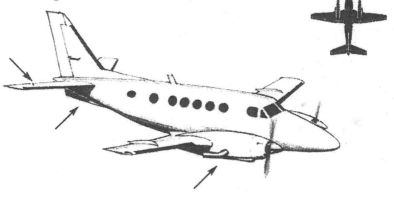

Beech Super King Air

Piper PA42 Cheyenne III, IV

Length: 43'5" (12.24 m) **Wingspan:** 47'8" (14.53 m) **Cruising speed:** 318 mph (512 km/h)

A business-size, low-wing twin turboprop with a T-tail, tip tanks, and rectangular windows. Typical Piper wing, strong fairing wing root to nacelle. (Compare with the Beech Super King Air [previous entry], which has optional tip tanks and round windows.)

Cheyennes first flew in 1980 and are exceptionally fast turboprop business planes. One circled the world in 1982 in 88 hrs. flying time, with 13 stops for fuel and rest. Executive seating for 6, less comfortable arrangements for up to 11 passengers. The newer Cheyenne IV or 400 (see silhouette) has engine nacelles that do not extend beyond the trailing edge of the wing.

Swearingen (Fairchild) Merlin II

Length: 40'1" (12.22 m) **Wingspan:** 45'11" (14 m) **Cruising speed:** 295 mph (475 km/h)

Small and fairly common. Low wing, conventional tail, turboprops. Resembles a smoother, bulkier, more streamlined Beech Queen Air. Three rather large rectangular windows on each side.

Swearingen, a company that specialized in putting turboprops, streamlined fairings, and pressurization into other companies' production aircraft, took the Queen Air wing and built a streamlined, pressurized fuselage for it from scratch. The small number of fairly large windows is unusual in a pressurized aircraft. Compare with the Beech King Air (five or six small windows) or the Cessna Conquest (six small windows; see page 118) for conventional treatment of similarly sized aircraft.

Swearingen (Fairchild) Merlin III, Fairchild 300

Length: 42'2" (12.85 m) **Wingspan:** 46'3" (14.1 m) **Cruising speed:** 288 mph (463 km/h)

Common. Combines low, symmetrically tapering wing with strongly swept tail plane mounted midway up and well forward on the tail fairing. Compare with larger Merlin IV (next entry). Similarly configured Handley Page Jetstream 31 has unswept tail plane mounted farther back on the fin and shows seven small round windows. The midtailed Rockwell Commander 700 has trapezoidal windows, unswept tail plane, and, unlike the Merlin or the Jetstream, no ventral fin.

A popular series of executive turboprops. Some early Merlin IIIs have only three or four windows to a side, and a variety of turboprop engines have been mounted on the same basic airframe. The strong dorsal and ventral fins shown on the Merlin and the Handley Page Jetstream are intended to improve handling when the plane is forced to fly on one engine. The 1984 Fairchild 300 has winglets.

Piper PA42 Cheyenne III

Swearingen Merlin II

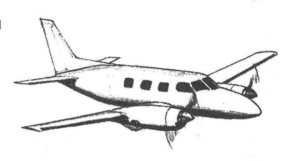

Fairchild Merlin IIIB

Fairchild Merlin IVA, Metro III, Fairchild 400
Length: 59'4" (18.08 m) **Wingspan:** 46'3" (14.1 m) **Cruising speed:** 279 mph (449 km/h)

Common. Combines low, symmetrically tapered wings and strongly swept tail plane mounted well forward on the tail fin fairing. Compare with the much smaller Merlin IIIB (previous entry).

Carrying 12 passengers in the Merlin IV executive cabin or up to 20 passengers in the Metro airliner cabin, this Swearingen-designed airplane has seen some use in the U.S. Midwest as a commuter airliner. It is quite rare as an executive plane. Some 300 delivered worldwide since 1971. The 1983 models introduced winglets; the Merlin IV was renamed Fairchild 400 in 1984. Later Merlin and Metro V have T-tail.

Beech 99 Airliner
Length: 44'7" (13.59 m) **Wingspan:** 45'10" (13.97 m) **Cruising speed:** 270 mph (434 km/h)

A common and variable aircraft. Combines low wing with two turboprop engines, conventional tail, and distinct ventral fin. Unfortunately, it has to be distinguished from similar planes, including its predecessor, the Beech Queen Air, by noting the window patterns. From the front, the 99s show one small rectangular window, five or six larger rectangular windows, the typical Beech gap on or opposite the passenger door, and a small oval window at the rear.

A couple hundred of the 15-passenger stretched and pressurized version of the Beech Queen Air are in service with dozens of small airlines. Built since 1965, with a couple of engine variations. A rather ordinary-looking aircraft, with a moderately swept tail fin (compared to the Queen Air) and a long, pointy nose.

Embraer EMB110 Bandeirante
Length: 47'10" (14.58 m) **Wingspan:** 50'3" (15.32 m) **Cruising speed:** 203 mph (327 km/h)

An uncommon commuter airliner. Low-winged, twin turboprops; in the air, a strong impression of rectangularity: Note the sharp extension of the tail fin down through the tail plane to a ventral fin; overhead, slightly tapering wing and tail planes look quite rectangular; engines with deep nacelles (to hold landing gear) extend very far forward of the wing. The wraparound cockpit windows are composed of eight separate planes, which is most unusual in recently built aircraft. The 1984 model has a dihedral in tail plane.

A 17- to 19-passenger unpressurized aircraft first delivered to the United States in 1976. The Bandeirante competes directly with such small commuter airliners as the Beech 99. The parent company, Empresa Brasilia de Aeronautica, builds single- and twin-engine Piper airplanes under license and a series of commuter jetliners.

Fairchild Metro III

Beech 99 Airliner

Embraer EMB110 Bandeirante

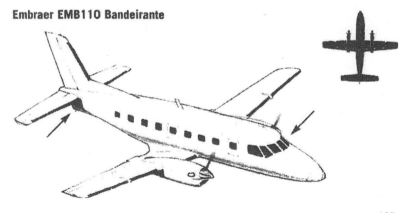

Embraer EMB120 Brasilia
Length: 64'5" (19.64 m) **Wingspan:** 74'10" (19.76 m) **Cruising speed:** 288 mph (463 km/h)

A 1984 introduction. **Very large low-wing twin turboprop with T-tail** (of twin T-tails, compare with the much smaller Piper Seminole, Duchess, Cheyenne III, and Beech Super King Air). The only other large twin T-tail is the **high-winged** de Havilland DHC8 Dash 8. High overhead, they might be confused if you do not pay attention to the wing placement.

Ordered by commuter airlines from coast to coast, this 30-passenger airliner includes state-of-the-art technology. The fuel-efficient Canadian-built turboprops have an unusual feature: fully disengageable propellers, so that the engines can be run at the loading gate. This feature allows not only passenger loading while the air conditioning and heating systems are kept on, as well as getting back in the air without delays associated with engine starting.

BAe Jetstream 31, Jetstream 32
Jetstream 32 data: Length: 47'1" (14.36 m) **Wingspan:** 52' (15.85 m)
Cruising speed: 288 mph (463 km/h)

A medium-sized commuter airliner: A square thing: **unswept wings and tail plane, which is mounted almost halfway up the tail plane, seven vertically oval passenger windows, multipieced windscreen.** Compared to the 31s, the 32 has an oversized belly bulge. The original had no belly fairing under the wing; **small but distinct ventral fin on 31s and 32s.**

A rare plane in North America, although it is locally very common in British Commonwealth nations, especially New Zealand, where it is a major short-hop aircraft. Designed by Handley Page and first flown in 1970, followed by the bankruptcy of Handley Page. BAe Systems did not change the basic size, but improved engines increased its cruising speed by some 20 mph (32 km/h).

Beech 1900 Airliner, 1900D
Length: 57'9" (17.60 m) **Wingspan:** 54'6" (16.61 m) **Cruising speed:** 280 mph (451 km/h)

Combines **low wing with T-tail fuselage-mounted stabilons just forward of tail;** typical Beech wing begins with **rectangular section from fuselage to engine; trailing edge tapers to tip more sharply than leading edge.**

A 19-passenger aircraft intended for commuter routes requiring frequent stops. The sharp dihedral in the low wing, combined with the T-tail, give the 1900 a unique appearance in the landing and takeoff pattern. Note also the very large double engine exhausts. The 1900D is a "tall body" with stand-up headroom.

Embraer EMB120 Brasilia

BAe Jetstream 31

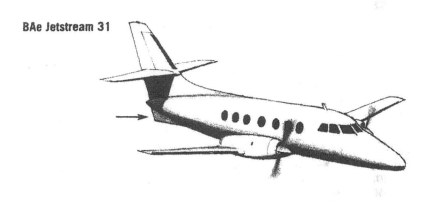

Beech 1900 Airliner

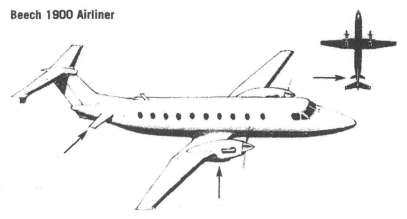

Saab-Scania 340A Commuter

Length: 63'9" (19.43 m) **Wingspan:** 70'4" (21.44 m) **Cruising speed:** estimated, 300 mph (483 km/h)

A fairly conventional-looking airplane: Tall, swept tail fin, strongly dihedral tail plane; deep fuselage is carried full depth well aft; unusual engine nacelles, which are narrow and deep, rise high above and show well below wing.

A 34-passenger airliner with wings and tail by Fairchild, the rest by Saab; assembled in Sweden. The aspect of the plane is unique: Strong dihedrals in tail planes tend to be unusually noticeable, as on the old Martin 404. The bulky body and slim wing will attract attention.

Grumman American G159 Gulfstream I

Length: 64'8" (19.72 m) **Wingspan:** 78'4" (23.88 m) **Cruising speed:** 288 mph (463 km/h)

Not common. Slim winged, short nosed; distinct swelling under engine nacelles houses landing gear. A stretched version, the G159 1C, is 10 ft. longer and shows seven rather than five oval passenger windows.

Carrying 24 passengers in the short version or 37 in the model 1C stretch, some 200 of these durable, but not particularly fuel-efficient, corporate planes operate in North America. Although built from 1960, with the stretching done in the early 1980s, they're not currently competitive with newer aircraft of the same capacity.

de Havilland DH104 Dove, Riley Turbo-Exec Dove

Length: 39'4" (12 m) **Wingspan:** 57' (17.37 m) **Cruising speed:** 162 mph (261 km/h)

Extraordinarily rare. Long, tapering wings; engines mounted well forward on the wing; distinctive bump over cockpit gives crew stand-up headroom. Originals show a conventional curved tail, whereas Riley turbocharged conversions have a swept, angular tail fin.

About 600 built by de Havilland between 1946 and 1968, many as military light transports. They became a popular executive aircraft after WWII, and the turbo conversions continue to fly in general aviation. A Dove with the old Gipsy Queen engines is a real rarity in North America. The first one you see is likely to be the last one.

128

Saab-Scania 340A Commuter

Grumman American G159 Gulfstream I

de Havilland DH104 Dove

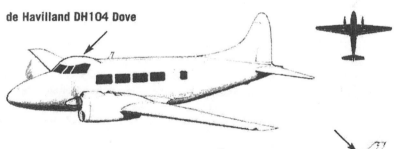

Riley Conversion

British Aerospace 748

Length: 67' (20.42 m) **Wingspan:** 102'5" (31.22 m) **Cruising speed:** 281 mph (452 km/h)

One of two modern **twin turboprops** that share the characteristic of **massive bulges on the bottom of the nacelles** (to house landing gear). Compare with the Japanese NAMC YS11, below. The BAe 748 has **strong wing dihedral, beginning at fuselage,** combined with **horizontal tail planes.** Convair 640 has similar wing and tail configuration but without the massive landing gear fairings. Passenger BAe 748s have 10 large, rounded windows. The NAMC has many small, square windows.

A stretched 748 ATP has much less bulbous engine/landing gear nacelles and 26 windows on its much longer fuselage, 85'4" (26 m) overall.

Convair CV240, 340, 440, 540, 580, 600, 640

CV580 data: Length: 81'6" (24.84 m) **Wingspan:** 105'4" (32.11 m)
Cruising speed: 300 mph (483 km/h)

Rare. A variety of highly similar **twin-engine, low-wing** airliners, with **slight dihedral in wing and horizontal tail planes.** In the United States and Canada, most are **turboprop conversions,** series 540 to 640. (CV580 is the most common.) Except for the engine nacelles, very similar to the BAe HS748 and NAMC YS11. Whether old piston or new turboprop, **the nacelles are slim** compared to the bulging, landing gear–holding nacelles on the HS748 and YS11.

The original 240, 340, 440 series, seating 40 to 50 passengers, with Pratt and Whitney radials, have been supplanted for the most part by turboprop conversions. A few made-from-scratch turboprops produced by Canadair— the Canadair CC-109—stayed in service as troop carriers in the Canadian armed forces until the 1990s. Model numbers reflect little except the time of manufacture or re-engining. However, the 340 and 440 were slightly stretched versions of the original 240.

NAMC YS11

Length: 86'3" (26.3 m) **Wingspan:** 104'11" (32 m) **Cruising speed:** 281 mph (452 km/h)

Rare but seen occasionally in Alaska and in the southwestern United States. **Massive landing gear fairings under nacelles** (compare with the BAe 748), **slight dihedral wing; horizontal tail plane; dozens of tiny rectangular windows.**

Either the limits of conventional airplane design were reached in the 1950s, or this is a virtual copy of the British Aerospace 748. Its design was begun in 1960, a year after the 748 went to the drawing board. The YS11 does carry 60 passengers, not 44, but is otherwise highly similar to the BAe 748; the windows are the most obvious difference.

BAe 748

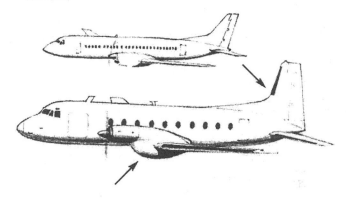

Convair 580

NAMC YS11

Curtiss C-46 Commando
Length: 76'4" (23.27 m) **Wingspan:** 108' (32.92 m) **Cruising speed:** 235 mph (378 km/h)

A very rare survivor. (Before deciding, make sure it's not a DC3.) **The plane with no nose; greenhouse cockpit windows.** The wings are like the DC3's: strongly tapered on the leading edge, straight on the trailing edge. Unlike the DC3, this plane has **fully retractable landing gear** and is larger, bulkier than the DC3.

Developed as a 36-passenger airliner in 1940 to compete with the DC3, it was built only as a military transport. A few dozen still survive with small, poor regional airlines; likeliest to be seen in the Caribbean, southwestern Alaska, along the Mexican border. It was never as common as the somewhat similar DC3.

Douglas DC3, C-47, Dakota
Length: 64'5" (19.65 m) **Wingspan:** 95' (28.96 m) **Cruising speed:** 194 mph (312 km/h)

Not common but widely distributed. A **tail dragger** that sits nose up on the flight line; in the air, **very short-nosed look,** as the wings are set well forward, and the large radials flank the cockpit area; wing tapers on the leading edge only; **tires of forward landing gear do not retract out of sight; tail wheel is nonretractable.**

First built in 1935 and flown the world over, with a few hundred surviving long after the assembly shut down in 1946. Seated 36 in unpressurized discomfort and as many as 50 in its troop-carrying configuration. Still flying passengers in North America. Scores parked on airfields and making occasional unscheduled freight trips. As with many aircraft with partially retractable wheels, the purpose is to allow for a relatively safe landing in the event that the gear is not, or cannot be, lowered.

Curtiss C-46 Commando

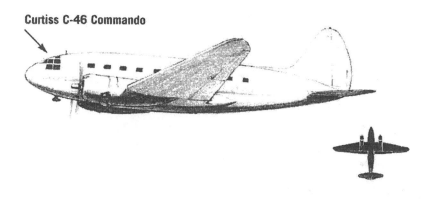

Douglas DC3

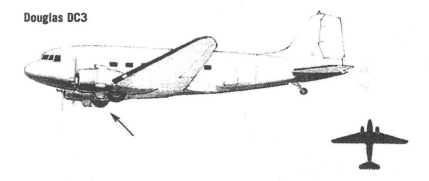

Beech 18, C-45

Length: 35'2" (10.72 m) **Wingspan:** 49'8" (15.14 m) **Cruising speed:** 185 mph (298 km/h)

Fading away but still flying; highly variable. Twin-engine, low-wing, distinctive Beech twin tail: Note that **tail plane does not extend through fins.** Seen with rounded (early) and squared-off (late model) wingtips.

The durable Beech 18 was built from 1937 to 1972, with thousands in WWII as C-45s. It has been refitted in a bewildering variety of forms: with tricycle gear to replace the semiretractable tail-dragging gear, in stretched versions, in long-nosed models, with turboprop engines, with conventional rather than double-fin tails, and, in one bizarre case, with a T-tail. **The odd window pattern—a long, rectangular passenger window surrounded by two smaller square windows—is always a good field mark.** The last production 18s were sold to Japan Airlines.

Lockheed 10, and 12 "Electra Jr."

Model 12 data: Length: 36'4" (11.07 m) **Wingspan:** 49'6" (15.09 m)
Cruising speed: 206 mph (331 km/h)

Very rare. These are similar, but the model 10 has five side windows; the model 12, three. **Twin radial engines on low-wing, classic double-fin Lockheed tail; tail plane extends through the fin; main landing gear quite visible when retracted into open wheel wells.**

The model 10, first flown in 1934, was America's first all-metal-skin airplane. Quickly adopted by airlines, it carried 12 passengers and a crew of 2; the smaller "Electra Jr." model 12, carrying 6 passengers and a crew of 2, was intended for the corporate plane and feeder-airline business. Although only a couple 12s and not more than five model 10s are flying, we could not exclude these grandparents of a famous family of propeller airliners, culminating in the Super Constellation (page 154).

Lockheed L18 Lodestar, C-60, PV-1, PV-2

Length: 49'10" (15.37 m) **Wingspan:** 65'6" (20.21 m) **Cruising speed:** 229 mph (368 km/h)

Rare and worth looking for. **Wing mounted just below midpoint of fuselage; twin tail; tail plane extends through tail fins; two radial engines.** The more common Beech 18 is much smaller and does not have the Lockheed-type tail planes extending through the vertical fins.

The premier short-haul airliner just as WWII started and a common personnel carrier (C-60) through the war. A distinctly tail-dragging aircraft, with the nose pointed up as if it should be flying, it's usually seen sitting idle on a runway apron. Carried 14 passengers in relative comfort, including a full lavatory in the rear of the aircraft. PV-1 and PV-2 were early WWII long-range patrol bombers.

Beech 18, C-45

Lockheed 12

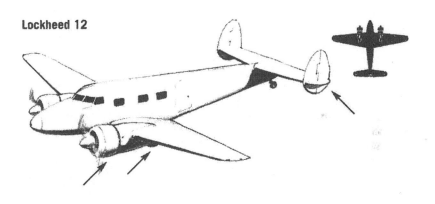

Lockheed L18 Lodestar

135

Cessna Bobcat, Crane, T-50, AT-8, C-78

Length: 32'9" (10 m) **Wingspan:** 41'11" (12.8 m) **Cruising speed:** 165 mph (265 km/h)

Rare, small, and old-fashioned-looking twin, with **huge radials compared to the size of the plane; long nosed, but the nose barely extends past the engine nacelles; partially retractable landing gear.**

Built by the thousands from 1940 to 1945 as a primary (T-50) and advanced (AT-8) multiengine trainer for the U.S. (Bobcat) and Canadian (Crane) armed forces. Several hundred served as light transports (C-78). Many converted to civil air after WWII, but wooden wings did not allow conversion to more efficient turboprops. Slightly underpowered, they're not really flyable on one engine; nevertheless, a durable, reliable short-haul aircraft.

North American B-25 Mitchell

Length: 52'11" (16.33 m) **Wingspan:** 67'7" (20.86 m) **Cruising speed:** 250 mph (402 km/h)

Very rare, variable. Combines **midwing with double tail fins.** Note that it is a high midwing and that the tail plane does not extend through the vertical fins. Compare with the somewhat similar Lockheed Lodestar, with its much lower wing mounting and tail plane extending through the twin tail fins.

Designed before WWII, more than 10,000 were built; losses kept the inventory to about 2,600 maximum during the war. Produced with and without the glass bombardier's nose; civil conversions usually have closed-in noses, and some have tip tanks; a few have passenger windows. Once fairly popular as an aerial sprayer. Carrier-launched B-25s made the token attack on Tokyo in April 1942; B-25s were the aircraft seen in the 1970s movie *Catch-22*.

Douglas A-26 Invader

Length: 53'10" (16.4 m) **Wingspan:** 70' (21.34 m) **Cruising speed:** 325 mph (523 km/h)

Very rare, variable, probably parked permanently. Look for the constants. **Wing mounted higher than the B-25's but not above fuselage; two huge, cylindrical engine nacelles extend well forward and back of the wing; nacelles mounted low on wing; long, bulging nose; shallow cockpit windows.**

Once you get the configuration, you can ignore the dozens of variations of the basic aircraft: A high-speed, large-capacity executive conversion may have completely enclosed nose, passenger windows, and tip tanks on the wings, but the basic wing and engine conformation is undisturbed and unmistakable. Known as the A-26 (for attack bomber) through WWII but redesignated B-26 after the war. The WWII B-26 was the Martin Marauder, with short, tapering engine nacelles.

Cessna Bobcat

North American B-25 Mitchell

Douglas A-26 Invader

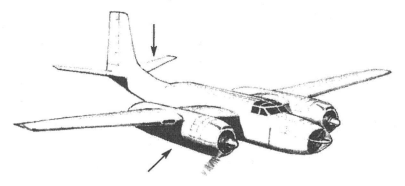

HIGH-WING TWINS

Partenavia P-68C
Length: 31'4" (9.55 m) **Wingspan:** 39'5" (12 m) **Cruising speed:** 185 mph (298 km/h)

Rare in North America. A very sleek and long-nosed high-wing twin with fixed gear and wheel fairings. Seen directly overhead, it could be confused with the much larger de Havilland Twin Otter, as both have constant-chord wings and tail planes. But note the Partenavia's unusual bracing fillet from fuselage to leading edge of tail plane/horizontal stabilizer.

The remaining Partenavias (fewer than 50 at this writing) in North America are generally used by flight schools as dual-control twin-engine trainers. Many more in Italy, where a bubble-nosed version is a police and search-and-rescue vehicle.

Mitsubishi MU2 Marquise, Solitaire
Marquise data: Length: 39'5" (12.01 m) **Wingspan:** 39'2" (11.94 m)
Cruising speed: Marquise, 340 mph (547 km/h); Solitaire, 370 mph (595 km/h)

Not common, small, high-wing twin turboprop. Tip tanks; tail plane set noticeably lower than wings. Earlier Japanese-built Marquise has bulging fuselage fairings to hold retractable wheels. American-assembled Solitaire has smooth fuselage into which gear retracts. Early Japanese Marquise models are 33 ft. long, as are all American Solitaires.

A moderately popular corporate plane. The relatively high cruising speed, combined with fuel efficiency and room for four to nine passengers, made it the hot-rod of twin turbos. It even became a popular plane to steal and use in the Caribbean drug-smuggling world. Several models (the plane comes with a variety of engines) have ranges up to 1,680 mi. (2,700 km), which is long for the class.

Partenavia P-68C

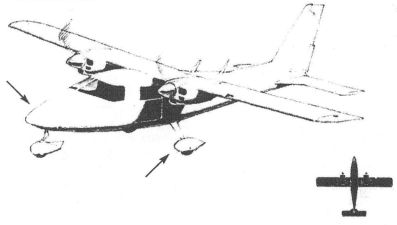

Mitsubishi MU2 Marquise

Gulfstream and Rockwell Commander, Shrike Commander, Aero Commander, etc.

Aero Commander 520 data: Length: 34'6" (10.52 m) **Wingspan:** 44'7" (13.6 m)
Cruising speed: 197 mph (317 km/h)
Turbo Commander 690 data: Length: 44'4" (13.51 m) **Wingspan:** 46'8" (14.22 m)
Cruising speed: 288 mph (463 km/h)
Shrike Commander (Aero Commander 500U) data: Length: 35'1" (10.69 m)
Wingspan: 49'2" (15 m) **Cruising speed:** 201 mph (323 km/h)

A complex family of airplanes, begun in 1948 with the four-passenger piston-engine Aero Commander, and proceeding through the turboprop Rockwell 690 and Gulfstream 840, 900, 980 and 1000, 1200 series, carrying as many as 10 passengers. All share certain characteristics: **high wing with slight dihedral, twin engines, strong dihedral in tail planes.** Models with turboprops from 690B on have small winglets. Very earliest four-passenger Aero Commanders and Shrike Commanders have a curved leading edge to the tail fin; all later models, a straight-edged, strongly swept tail fin. Another characteristic, from the Aero Commander on, is the **upswept fuselage,** which becomes increasingly distinct as the later models appear. Long nosed and streamlined, compared to other high-wing twins. The streamlining effect is visually enhanced by the dihedrals in wing and tail plane. The authors accept the judgment of other airplane aficionados, who lump the whole, varied, 25-year-old class of airplanes under the single category: Commanders.

de Havilland DHC6 Twin Otter

Length: 51'9" (15.77 m) **Wingspan:** 65' (19.81 m) **Cruising speed:** 200 mph (322 km/h)

Slim bodied, with long, thin, high wings and twin turboprops; fixed gear; conventional tail; wing braced from fuselage at landing gear root. Compare with somewhat similar and much rarer GAF Nomad (page 142), whose wing brace rises from the landing gear itself.

Built from 1965 to 1988, it's one of the most popular small airline and air-taxi planes ever flown. More than 300 are in service. Carries 14 to 18 passengers in a fairly quiet, center-aisle cabin. Very short takeoff and landing qualities; can take off across the width of most airports. Seen as a float plane, though not as often as the de Havilland single-engine Otter.

Gulfstream Commander 900

Rockwell Turbo 690

Aero Commander 680

Aero Commander 520

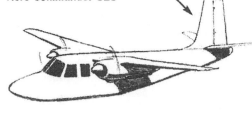

de Havilland DHC6 Twin Otter

Pilatus Britten-Norman Islander BN-2, Trislander MK III

Islander data: Length: 35'8" (10.87 m) **Wingspan:** 49' (14.94 m) **Cruising speed:** 150 mph (241 km/h)

A plane of odd geometry. **Fuselage rectangular in cross-section; varied window shapes:** rectangular, trapezoidal, rhomboid; **Hershey bar wing and tail; curved wingtips are auxiliary fuel tanks; double wheels** on **lumpy, nonretractable landing gear.**

Designed for fuel-efficient, low-speed, low-density commuter routes. The earlier versions had a short nose, whereas the last version, the Trislander, has a longer fuselage, a T-tail, and a third engine mounted high on the tail fin. A low-technology airplane, it has been manufactured under license in Romania and assembled from supplied parts in the Philippines and the United States. Seats up to 18 passengers and a single pilot; no aisle, entry through doors directly to seats.

GAF (Government Aircraft Factory, Australia) Nomad

Length: 41'2" (12.56 m); long-nosed model N24, 47'1" (14.36 m) **Wingspan:** 54'2" (16.51 m)
Cruising speed: 193 mph (311 km/h)

Rare in North America. **High wing, twin turboprops; tail plane mounted partway up tail fin; wing struts rise out of the wheel pants of the fixed landing gear** (compare with the de Havilland Twin Otter strut and tail).

Developed by the Australian factory as a military search-and-rescue and light-transport plane in 1971. Two civil versions: the short-nosed N22 for 12 passengers, the long-nosed N24 for 15. Competitive in the same market as the DHC Twin Otter and, as such, may be seen fitted with floats. Several were ordered by North American air-taxis and commuter airlines. The U.S. Customs and Immigration Enforcement Service flies several along the Gulf Coast.

CASA C212 Aviocar

Length: 49'10" (15.2 m) **Wingspan:** 62'4" (19 m) **Cruising speed:** 196 mph (315 km/h)

Still rare. **Stubby look, high wings, twin turborprops, upswept rear fuselage, conventional tail, nonretractable gear.** Compare with equally stubby Shorts Skyvan (next entry), which has braced wing and upswept fuselage, or de Havilland Dash 8 (page 146), which is upswept but has T-tail and retractable gear.

CASA is Spain's aircraft manufacturer, and the Aviocar is its design. Originally, a 16-person paratroop transport and utility freighter or air ambulance. The civil versions can carry 19 passengers and operate from the shortest and roughest airstrips. A popular commuter aircraft in the Far East and African countries, where it replaces the aging WWII-surplus planes that have ended their careers in Third World airlines.

Pilatus Britten-Norman Islander

Pilatus Britten-Norman Trislander

GAF Nomad

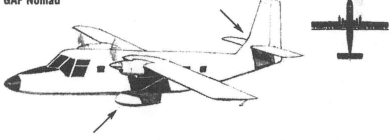

CASA C212 Aviocar

Shorts Skyliner, Skyvan

Length: 40'1" (12.22 m) **Wingspan:** 64'11" (19.79 m) **Cruising speed:** 173 mph (278 km/h)

Stubby, fixed landing gear with wheels tucked up under body; twin tail fins; long, thin wings with braces.

Resembling a flying bathtub with a thin wing glued on top, the Short Brothers Skyvans serve small airlines in eastern North America and Alaska. The plane, built of a metal-resin composite with little or no insulation, seems remarkably noisy to passengers who took to flying after the DC3 era. More than 150 Skyvans (or more luxuriously appointed Skyliners) were built from 1964 to 1982.

Shorts 330, C-23 Sherpa, 360

Length: 58' (17.69 m) **Wingspan:** 74'8" (22.67 m) **Cruising speed:** 173 mph (278 km/h)

Bizarre configuration: long, thin, untapered wing with large strut; semiretractable wheels show even in flight. Most models have a double tail fin, like their brother, the Shorts Skyvan (preceding entry). One is not surprised that the builder, Short Brothers Company, was once a leading manufacturer of flying boats.

Introduced in 1976 as a fuel-efficient feeder airliner, the very lightweight and low-maintenance 30-seat Shorts 330 is of composite metal and resin construction. A slightly larger version, the Shorts 360, carrying 36 passengers and bearing a conventional tail, has been purchased by several North American commuter airlines. A variant 330 is used by U.S. forces in Europe for shuttling between bases, rapidly declining in numbers.

Dornier 228, Series 200

Length: 54'4" (16.56 m) **Wingspan:** 55'8" (16.97 m) **Cruising speed:** 266 mph (428 km/h)

Not common but heavily used as a feeder airline, so commonly seen. Long nosed, the high-tech wing shape somewhat resembles the Islander/Trislander series but with very short engine nacelles; boxy, bumpy fuselage shape.

If you set out to build a small, slow, durable, economical short-haul airliner and didn't much care what it looked like, you'd get a Dornier. Travelers to Europe and Africa will see an occasional 228 in military camouflage or dressed up with radar pods for marine surveillance.

Shorts Skyvan

Shorts 330

Shorts 360

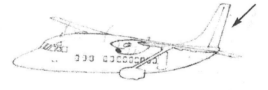

Dornier 228, Series 200

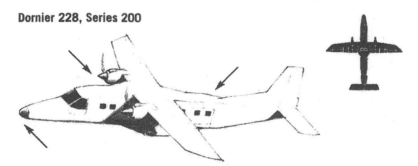

145

Aerospatiale (Nord) 262, Mohawk 298

Length: 63'3" (19.28 m) **Wingspan:** 71'10" (21.9 m) **Cruising speed:** 233 mph (375 km/h)

Rare, local. **High, thin, tapering wings; bulging landing gear nacelles on fuselage; tires exposed even when retracted.**

This 26-passenger short-haul airliner went into service in 1963 and, with improved engines, survived into the 1980s. It was one of the first of the high-efficiency, short-distance airliners and was soon surpassed by later models (the Shorts 300, for example). Only 110 were built; perhaps a few still carry passengers.

de Havilland Dash 8, Bombardier Q100, Q300, and Q400

Q100 data: Length: 73' (22.3 m) **Wingspan:** 85' (25.9 m) **Cruising speed:** 308 mph (496 km/h)

A family of **high-wing, T-tailed,** turboprop airliners that began with the Dash 8, a smaller version of the DeHavilland four-engine Dash 7 (page 154). The Q100 and the essentially identical Dash 8 are the only large high-winged T-tails; the Q300, somewhat larger still; and the Q400, extremely large. The Q400, at 107'9" (32.84 m), is fully half again as long as the original Dash 8/Q100. The odd fuselage on the Q400, showing **a long blank area between the first passenger window and the cockpit window,** occurs because the baggage for the 80-passenger Q400 must be balanced fore and aft; thus, the **two baggage doors on the Q400's starboard side.**

Stretched jetliners are common enough, but propeller-driven airframes usually stay the same through their years. The Q series, even including the 400, has the great attraction of using rather short runways, an ability to make an approach at a relatively steep angle, and very quiet engines. This makes them popular for travel to and over densely settled areas. By far the largest number of Q400s have been sold to European air carriers. With a cruising speed of 400 mph (648 km/h) and a quick ascent to its 25,000 ft. (7,620 m) cruising altitude, it is as fast as jetliners on routes of fewer than 500 mi. (805 km).

Aerospatiale 262

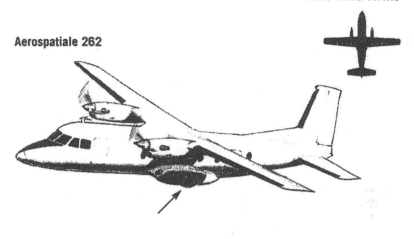

de Havilland Dash 8

Bombardier Q400

Aerospatiale/Aeritalia ATR 42, ATR 72
ATR 42 data: Length: 74'5" (22.67 m) **Wingspan:** 80'8" (24.57 m)
Cruising speed: 279 mph (450 km/h)

Few in number but continually in use as airline feeder, so visually common. Huge, sweeping tail fin to a high cross, almost T-tail; fuselage tapers to a tail cone (a close-out fairing); very large landing gear fairing under fuselage.

Roughly the same size and configuration as a de Havilland Dash 8, the ATR is heavier looking, thanks to the landing gear fairings and tail fin. A newer model, the ATR 72, is a very large high-wing twin, a 15 ft. (4.57 m) stretch of the ATR 42, but visual outlines and shapes are identical.

Fokker F27, Fokker 50 Friendship
Fokker 50 data: Length: 82'10" (25.25 m) **Wingspan:** 95'2" (29 m)
Cruising speed: 311 mph (500 km/h)

Not as long nosed as the Dornier and much larger. The Fokker F27 (illustrated) is typical: **Long fairing to tail parallels the upsweep of the rear fuselage; pointy-nosed long engine nacelles extend symmetrically ahead and behind wing.**

The new Fokker 50 is a higher-performance, modern-materials version of the old F27 and stretch F227. If you catch a 50 sitting on the runway, engines off, it will show a six-bladed propeller, and its windows are more TV-screen-shaped than the oval ones on the F27 and F227. A fairly common feeder airliner.

Fairchild C-119 Flying Boxcar
Length: 89'5" (27.25 m) **Wingspan:** 109'3" (33.3 m) **Cruising speed:** 200 mph (km/h)

Rare and probably parked; a bathtub body with a twin-boom tail.

None in military service. If you're watching WWII newsreels and you think you see a C-119, it is probably the predecessor C-82 Packet. C-119s were heavily used during the Korean War, and some in Vietnam, mostly as gunships. Just to confuse things, a few C-119s with jet-assisted takeoff were called C-119 Packets. Very few hours flown today, but many are parked around the United States.

Aerospatiale/Aeritalia ATR 42

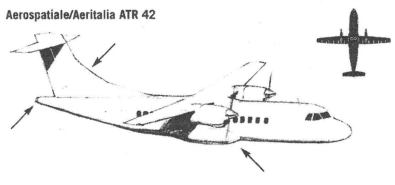

Fokker F27

Fairchild C-119 Flying Boxcar

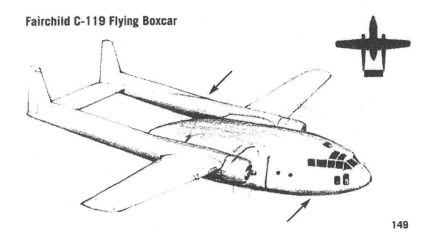

TWIN-BOOM AND CANARD TWINS

Cessna Skymaster 337, O-2

Length: 29'9" (9.07 m) **Wingspan:** 38'2" (11.63 m) **Cruising speed:** 173 mph (278 km/h)

Fairly common. One of two smallish twin-boom planes you'll see; compare with the military-only OV-10, page 230. The combination of twin booms and in-line engines—one pushing, one pulling—is unique.

More than 1,200 337s fly in the United States and Canada. The original idea was to build a plane with twin-engine redundant safety that, in the case of the failure of one engine, would be simple to fly because the pilot would not have to compensate for the sudden and persistent torque of a wing-mounted, off-center engine. Cessna hoped that the U.S. government would permit single-engine-rated pilots to fly the 337. That was not allowed, and production stopped in 1980 except for a few, named the Reims Milirole, built in France. As the O-2, more than 400 were in U.S. Army service.

Adam Aircraft A500, A700

A500 data: Length: 36'8" (11.18 m) **Wingspan:** 44' (13.4 m) **Cruising speed:** 265 mph (426 km/h)

A remarkable-looking twin-boom small plane (compare with the Cessna Skymaster, previous entry; OV-10 Bronco, page 230) with the horizontal tail surface at the top of the tail fins (like the Bronco), three small oval windows, and large turboprop air intakes in the cowling and sides of the rear fuselage.

The all-carbon-fiber A500 began production in 2004 while the larger jet engine A700 was at least temporarily in abeyance. The company began building a few A700s in hopes of receiving FAA certification in 2005 or later. The A500, with an airborne hang time of 7 hrs. and a range of 1,324 mi. (2,130 km), has attracted orders from contractors that provide surveillance for the U.S. government. The slightly longer fuselage on the jet engine 700 makes room for a toilet.

Cessna Skymaster 337

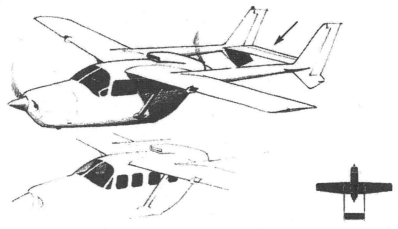

Skymaster, O-2

Adam Aircraft A500, A700

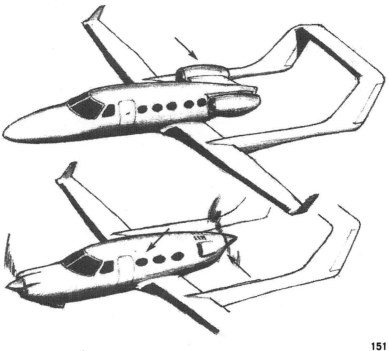

151

Piaggio P180 Avanti
Length: 46'6" (14.17 m) **Wingspan:** 45'5" (13.84 m) **Forewing span:** 10'9" (3.28 m)
Cruising speed: 368 mph (593 km/h)

If it weren't for the **short canard wing under the cockpit** and the **upside-down and backward-pushing turboprop engines,** this T-tailed metal-skinned aircraft would look fairly conventional.

This is a tweaked-up airplane, using a standard Piaggio wing design but inserting it into the middle of the rear end of the fuselage, taking popular turboprop engines but reversing them to pushers, and taking metal skin but shaping it in large sections and conforming the interior structure to the skin, the reverse of normal manufacturing. In the air, with its way-back wing, it looks as if the fuselage is dragging the rest of the plane along behind it.

Beech Starship I
Length: 46'1" (14.04 m) **Wingspan:** 54'5" (16.6 m) **Forewing span, extended:** 25'6" (7.79 m)
Cruising speed: 340 mph (546 km/h)

With its **tall (8'1", 2.45 m) winglets on a swept-back rear-mounted wing** and its **low, variable-angle forewing** and factory-delivered **pure white paint job,** this pusher-prop is highly noticeable.

This was the first of the new-generation pusher-props to be delivered to customers. Although few were produced, Starships have been seen across the United States and Canada. Rare but **visually obvious and unforgettable,** like a bald eagle.

Piaggio P180 Avanti

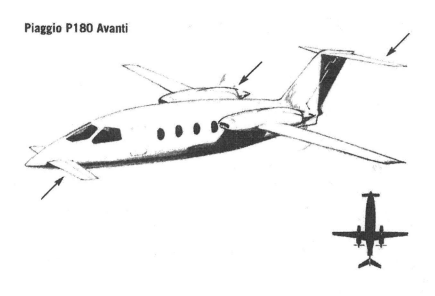

Beech Starship I

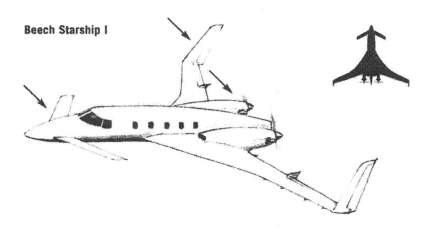

FOUR-ENGINE PROPS

de Havilland DHC7 Dash 7
Length: 80'8" (24.58 m) **Wingspan:** 93' (28.35 m) **Cruising speed:** 235 mph (378 km/h) Mach 0.354

Common. The only four-engine, high-wing, T-tail commercial aircraft in North America. Even when seen directly overhead, when it might be confused with the high-wing, conventional-tail C-130 Hercules, it is much slimmer and combines four engines with nacelles that do not show behind the wing, with a symmetrical taper on both edges of the wing from the fuselage to the wingtip.

A popular short-haul airliner, this Canadian import can carry 50 passengers from rural airports with very short runways. A few windowless models are used for air-freight operations, mostly in the Canadian backcountry. The Canadian Coast Guard flies a marine reconnaissance type (the DHC-7R Ranger) with bubble observer windows on the lower part of the fuselage and a belly-bulge radar dome.

Lockheed Constellation (C-69, C-121)
L1049 Super Constellation data: Length: 116'2" (35.41 m) **Wingspan:** 123' (37.5 m)
Cruising speed: 260 mph (418 km/h)

Rare, with almost none still flying. A very large four-engine, low-wing airliner/air-cargo hauler with triple tail fins; tail plane extends through outboard fins.

Once the queen of the transoceanic airways, a few Connies rest on runway aprons between charter flights. Most common was the L1049, carrying up to 110 passengers, built from 1943 to 1958. A few were converted to radar planes, designated EC-121, U.S. Air Force, and Navy. These had top and bottom radar bulges at the wing area of the fuselage. The rarest is the last model, the L1649, with a wing design similar to the Electra/Orion's: a straight leading edge perpendicular to the centerline of the fuselage.

Vickers Viscount 700
Length: 81'2" (24.75 m) **Wingspan:** 94' (28.66 m) **Cruising speed:** 315 mph (507 km/h)

A large, rare curiosity, four-turboprop airliner with rather large oval passenger windows: bumpy cockpit with an odd, shouldered effect (see the de Havilland Heron, next entry, for a similar treatment); very long, slim engine nacelles; three-piece cockpit side windows; slight dihedral in wing; sharp dihedral in tail plane.

First prototype flown in 1948; first production, 700 in 1952, carrying 40 to 59 passengers, depending on seating chosen. Originally named the Viceroy, after the title of the British ruler of India; renamed the Viscount after Indian independence. The world's first turboprop airliner, the Viscount managed to penetrate the U.S. market briefly in the late 1950s.

de Havilland DHC 7 Dash 7

Lockheed Constellation

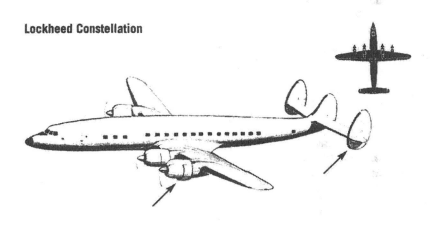

Vickers Viscount 700

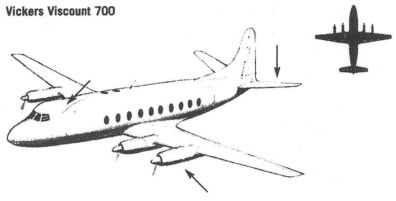

de Havilland Heron

Length: 48'6" (14.8 m) **Wingspan:** 71'6" (21.8 m) **Cruising speed:** 285 mph (459 km/h)

Very rare. Except for the **bulging bump over the cockpit**, a wonderfully symmetrical plane. **Slight dihedral in wings and tail planes;** overhead, **symmetrically tapering wing and tail surfaces.**

Popular airframes are hard to kill: The twin-engine British transport Dove was scaled up and given four engines to become the Heron. Several private companies have put turboprop engines on Herons, the most common a Riley Turbo Skyliner. Except as executive planes, you are most likely to encounter the few remaining Herons in the Caribbean. Note the classic British touch: Engines are centered vertically on the wing.

Douglas DC4, DC6, and DC7

(Old military designations: The DC4 was the C-54 Skymaster; the DC6, the C-118 Liftmaster)
Lengths: DC4 and DC6, 93'11" (28.6 m); DC6A and DC6B, 110'7" (30.66 m); DC7, 112'3" (34.21 m)
Wingspans: DC4, DC6, and DC7B, 117'6" (35.8 m); DC7C, 127'6" (38.86 m)
Cruising speeds: DC4, 227 mph (365 km/h); DC6, 313 mph (504 km/h); DC7, 310 mph (499 km/h)

Once you've positively identified one of the DC series, picking the specific one is a matter of size: **The only conventional-tail planes with four radial engines in nacelles that do not extend behind the wing's trailing edge.** (Constellation, page 154, has similar engine nacelles.) **DC4s have round windows; others are square.**

Now scarce as hens' teeth, the DC series, beginning with the pre-WWII DC4, once dominated U.S. aviation. All powered with radial piston engines, they became increasingly uneconomical in the face of new and sophisticated turboprop aircraft and did not survive well into the jet age. As military C-54 Skymasters, they ferried troops through the Korean War era. For the few remaining, separate them from other four-engine propeller jobs by the clearly radial piston engines. (Electras and CL-44s are turboprops, with slim, forward-extending engine housings; Herons have in-line piston engines that resemble four Spitfire or Mustang noses mounted on the wings, or they have been converted to turboprops.) The unstretched DC6 has no passenger windows forward of the wing; the DC6A and DC6B have two windows ahead of the wing; DC7s have three forward windows. The last and largest of the series, the DC7C, has the wingspan increased by 5 ft. on each side by the insertion of a rectangular 5 ft. wing root at the fuselage, a good mark when the craft is directly overhead. In general, overhead, the DC4, DC6, and DC7 series are marked by the engines' showing only forward of the leading edge and by the symmetrically tapering tail planes --the DC4 tail plane is rounded, much like an old Piper Cub's. (An Electra's leading wing edges make a straight line at right angles to the fuselage, and the tail plane edges are not symmetrical. Similar four-engine prop jobs show some nacelle behind the wing.)

de Havilland Heron

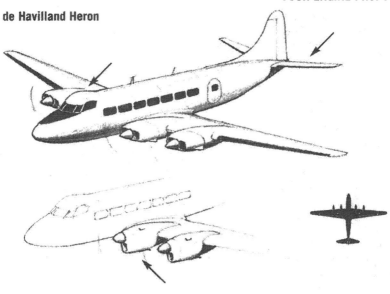

Douglas DC6

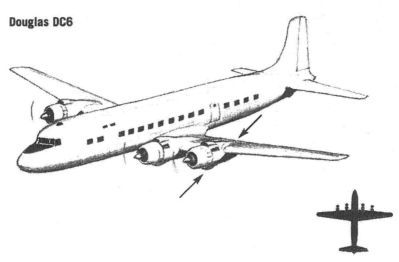

Lockheed L188 Electra
Length: 104'6" (31.8 m) **Wingspan:** 99' (30.18 m) **Cruising speed:** 405 mph (652 km/h)

Rare. Large, low wing, with four turboprops; leading edge of wing straight and at right angles to fuselage; conventional tail. Military and weather reconnaissance version, P-3 Orion, in limited use.

The jet-prop Electra came into service in 1959, just before the jet age, and in its first 18 months, its image was tarnished by two fatal crashes owing to structural problems in the wing. Buyer resistance lasted until the small, true-jet airliners had grabbed the commercial market. But the refitted Electras remain in service today as feeder airliners and especially as cargo planes. Like the newer CL44 and Dash 7, the turboprop Electra is much more fuel efficient than jet aircraft and operates at nearly 80% of jet speeds.

There is one possible confusion: Directly overhead, the plane resembles Lockheed's military C-130 Hercules; you may not see that the C-130 has a high wing and an upswept rear fuselage. Note the difference in the nose shapes of the C-130 and the L188. (See Lockheed P-3 Orion, page 208.)

Canadair CL44
Length: 151'10" (46.28 m) **Wingspan:** 142'3" (43.37 m) **Cruising speed:** 380 mph (611 km/h)

Four turboprops on midwings and ring around the tail where the fuselage swings open; fuselage hinged on port side; cockpit windows extend to top of fuselage: Compare with low-wing, radial-engine DC4 series; overhead, slim turboprop engine nacelles extend far forward of the wing's leading edge.

The CL44 is a fuel-efficient, long-range cargo plane, with a very few passenger versions in service in Canada. Except for the massive tail fin, it looks very conventional. First flown in 1959. The hinge area forward of the tail is usually painted a color different from the rest of the fuselage. Developed from the British Britannia as the CC-106 transport for the Canadian armed forces; then, with the swing tail, developed into the civilian CL44. CL44s are least uncommon at East Coast airports, where they haul freight on the North Atlantic routes.

Lockheed L188 Electra

Canadair CL44

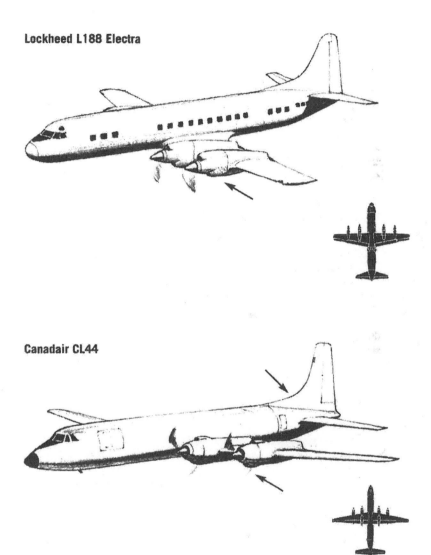

BUSINESS JETS

Cessna Citation Mustang
Length: 38'11" (11.86 m) **Wingspan:** 38'11" (12.87 m)
Cruising speed (estimated): 390 mph (630 km/h)

The newest Cessna, this quite small business jet carries four passengers. Note the three horizontal oval windows, resembling Gulfstream's style, quite unlike all its other Citations, with truncated rhomboids. Overhead, the long, thin wings are swept slightly on the leading edge only, and that edge enters the fuselate with a noticeable fairing or kink. The all-metal fuselage does not taper quickly to the tail; T-tail plane is horizontal and swept.

Cessna considers the $3,000,000 Mustang to be a reasonably priced upgrade from a similar-sized propeller-driven executive transporter. Although it has club-style seating for four and the ability to get above most weather, with an expected certified ceiling of 41,000 ft. (12,496 m), the Mustang, like other small business jets, lacks long-range executive amenities; there's no galley, and it has a small potty. It's a flying limousine.

Raytheon Premier I
Length: 46' (14.02 m) **Wingspan:** 44'6" (13.56 m) **Cruising speed:** 406 mph (645 km/h)

This is a unique small T-tailed jet with only three passenger windows each side, a recurved doorway that flows into the huge belly fairing enclosing the landing gear, wing roots, and mechanicals. The moderately swept wings have straight leading edges and kinked trailing edges. Overhead, the wings will look small and thin compared to the fuselage. Small engines in cowlings that conceal the jet exhaust tubes and thrust reversers.

In 2001, the Premier I was the first business jet to be FAA certified with an all-composite body (the wings are metal) molded in a series of subtle and complex curves to make it both bulky enough for comfort inside and aerodynamic on the outside.

Eclipse Aviation 500
Length: 33'1" (10.1 m) **Wingspan:** 37'5" (11.4 m) **Cruising speed:** 430 mph (694 km/h)

A very small twin jet, comparable in size to the average twin-prop general-aviation Cessnas or Pipers. Overhead, you may be able to see the rear wheels of the tricycle landing gear tucked up but uncovered. The bottle-shaped turbofan engines sit on angular mounts that reach to the very aft portion of the engine.

If the sky really does fill with mini-jets someday (typically carrying a pilot and three passengers), the Eclipse 500 will be as common as a yellow cab and be used as such. There are no luxuries, just relatively inexpensive transportation. More than three ft. shorter than Cessna's smallest craft, the Citation Mustang, this is likely to be the smallest twin-jet passenger plane ever built.

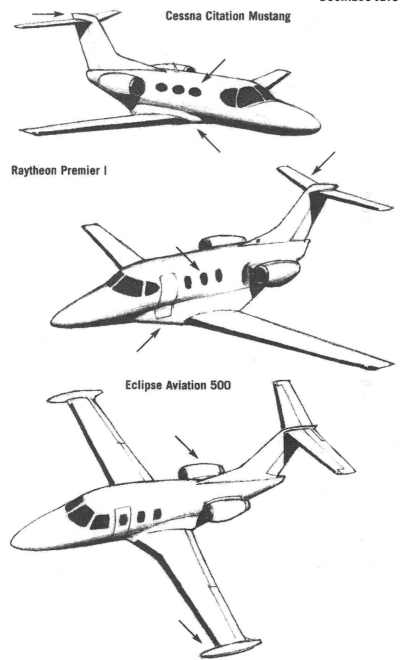

Cessna Citation Mustang

Raytheon Premier I

Eclipse Aviation 500

Gates Learjet 23, 24D

Model 24D data: Length: 43'3" (12.5 m) **Wingspan:** 35'7" (10.84 m)
Cruising speed: 481 mph (774 km/h)

The original small Learjet. Fuselage-mounted twin jets reach over the wing's trailing edge; tip tanks; wings with straight trailing edge; evenly tapered swept leading edge.

The four-passenger Learjet 23 and the six-passenger Learjet 24 are usually distinguished by the number of windows and the tail configuration. The 23 shows two passenger windows on the right side and one on the left, behind the passenger door. Most 24s show three passenger windows on the right and two on the left. Model 23s have a bullet at the center of the tail plane; most 24s do not.

Gates Learjet 25D, 28, 29

Length: 47'7" (14.5 m) **Wingspan:** 35'7" (10.85 m) **Cruising speed:** 528 mph (850 km/h)

One of a family of similar Learjets. The 25 series has five windows on the right and four on the left, behind the passenger door; wings have straight trailing edge; leading edge sweeps evenly (compare with the Learjet 35 or 36); T-tail.

The eight-passenger 25 is the stretched version of the successful Learjet series 23/24. At a quick glance, it could be confused with the larger Learjet 35 or 36, but note the 2 ft. equal-chord wing extension and the much larger engines on the 35 and 36. Models 28 and 29 use the Longhorn wing.

Learjet 35A, 36A

Length: 48'8" (14.8 m) **Wingspan:** 39'6" (12 m) **Cruising speed:** 529 mph (851 km/h)

Like the Learjet 25 but with large turbofan engines that extend above the top of the fuselage; wings lengthened by a 2 ft. equal-chord extension at the wingtip; five windows on the right, four on the left.

Introduced in 1973. Increased wingspan and larger engines make the 35 (eight-passenger) and 36 (luxury seating for four) capable of nonstop transcontinental or intercontinental range. Newest in the 30 series is the 31A, with the Longhorn wing.

Learjet, Bombardier 31A

Length: 48'9" (14.83 m) **Wingspan:** 43'10" (13.35 m) **Cruising speed:** 483 mph (778 km/h)

The 31A took the basic fuselage of the Learjet 35/36, at that time the largest Learjet, and added the Learjet 60 wing (lower sketch). Changes at Bombardier include the **highly visible strakes** and, at eye level or below, the **pairs of stall fences** on each wing (upper sketch). This is the economy version of the 60, with a smaller-diameter fuselage (read less headroom), slightly lower speeds, and shorter range. On the ground, the cockpit windshield shows **no small side window** as seen on the 60.

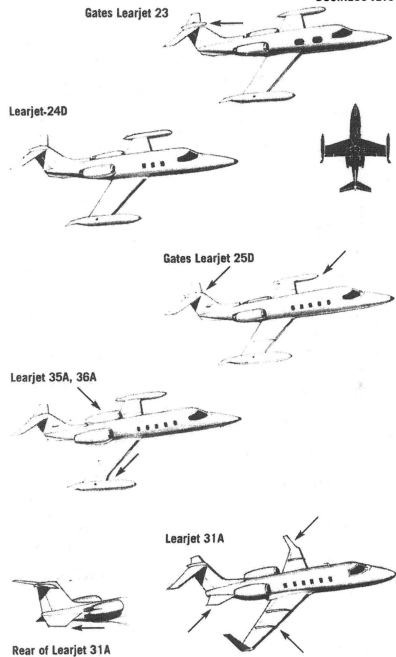

Gates Learjet 23

Learjet-24D

Gates Learjet 25D

Learjet 35A, 36A

Learjet 31A

Rear of Learjet 31A

Bombardier Learjet 40, 45, 45XR
45XR data: Length: 57'7" (17.56 m) **Wingspan:** 48' (14.56 m) **Cruising speed:** 526 mph (846 km/h)

These light jets are "clean sheet" new airplanes: faster, lighter (thus more load capacity), and roomier than the 41A and 60 models they will replace. Their **wings** continue the recent Learjet pattern, **swept on the leading edge only.** Unlike models 31A and 60, they have **clean wings, no stall fences.** Like all new Bombardier Learjets, they share large paired strakes. The **model 40** is slightly shorter than the 45, by about 3 ft., and **has seven rectangular side windows instead of eight.** The 40 and 45 have noticeably long noses and simple two-piece windshields.

Learjet 55, 60, and Bombardier Learjet 60
60-series data: Length: 58'5" (17.8 m) **Wingspan:** 43'9" (13.33 m)
Cruising speed: Bombardier model 60, 546 mph (880 km/h)

The Learjet 55 and 60 introduced what remains the standard wing format for subsequent Learjets; they are the only business jets with a swept forward edge and an orthogonal (right angle to the fuselage) trailing edge. Bombardier added more powerful engines and large distinct strakes (see sketch). The similarly sized and T-tailed Beechjet 400 has six oval, instead of five rectangular, windows on each side and a slight sweep to the wing's trailing edge. The slightly shorter and four-windowed Learjet 55, briefly called the Longhorn, was Learjet's 1981 entry into the medium-sized corporate-jet market.

Hawker 400XP (Raytheon), Beechjet 400A, Mitsubishi Diamond, T-1A Jayhawk
Length: 48'5" (14.75 m) **Wingspan:** 43'6" (13.25 m) **Cruising speed:** 483 mph (778 km/h)

One of the smallest of the T-tailed, swept-wing aircraft without tip tanks, and fuselage-mounted engines; compare with the much larger Canadair Challenger (page 180). The newest version, the Hawker 400XP (so "branded" by Raytheon), and the last ones sold as 400As have only five oval windows each side, dropping the rearmost window. The 400XP engines show short exhaust and clamshell thrust reversers; 400As have all-enclosing engine nacelles. The original 400A has six oval windows; the Canadair Challenger has six rectangular windows that begin behind the cabin door.

Beech acquired the Mitsubishi design and world rights (outside of Japan) in 1985. Sold in a modified form to the U.S. Air Force as the T-1A for pilot training, simulating in-flight refueling from the much larger tankers. This is the only American-built business jet that hasn't grown an inch through several model changes. The tweaked 400XP (extra payload) is capable of carrying one more passenger (or the equivalent in fuel) than the 400A. There are only so many ways to design a business jet; it's really convergent evolution.

Learjet 40, 45, 45XR

Learjet 55

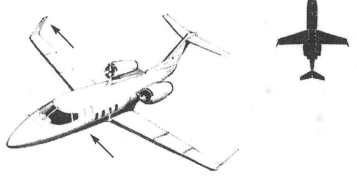

Beechjet 400A

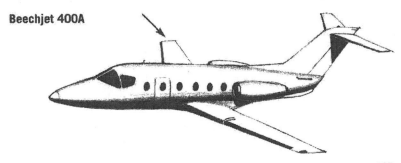

Cessna Citation I, II, SII, V, Citationjet
Citation I data: Length: 43'6" (13.26 m) **Wingspan:** 47'1" (14.35 m)
Cruising speed: 420 mph (675 km/h)

A family of conventional-looking **twin-fuselage-mounted jet** business aircraft; **unswept low tail plane, unswept wings tapering symmetrically.**

The first Cessna was the Fanjet 500, introduced in 1969. Citation I featured increased wingspan; Citation II is a stretch, with 5 ft. (1.52 m) more length and wingspan, and six windows instead of four. The high-performance SII has leading-edge extension at the wing root for more lift. The Citation V has the SII wing, stretches 2 ft. (0.61 m), and adds a window, for a total of seven. The Citationjet, similar to the old Citation I, is an economy version for the less affluent.

Cessna CJ1, CJ2, CJ3
CJ2 data: Length: 47'8" (14.53 m) **Wingspan:** 49'6" **Cruising speed:** 460 mph (740 km/h)

Another family of small jets from Cessna: **Swept tail fin rises abruptly without fairing; swept T-tail; most recent models with a new tail cone "bullet." Tapered, unswept wings, no winglets.** Distinguishable from the Citations mainly by the tail configuration.

Another Cessna line emphasizing short-runway capacity, economy, and quiet. The latest CJ3, just 3 ft. longer in span and length than the CJ2, will compete directly with the new Citation Bravo for range, speed, and interior space.

Cessna Encore (Ultra), Bravo
Length: 48'11" (14.9 m) **Wingspan:** 54'1" (16.49 m) **Cruising speed:** 494 mph (795 km/h)

Cessna makes three business jets with **low-mounted dihedral tail fins:** The Encore is the smallest, with just **seven of Cessna's unique and diagnostic passenger windows, a sort of truncated diamond, including one in the passenger door.** Unlike the similar Cessna Encore and Cessna Sovereign, **the wing root is well aft of the passenger door, with a modest fairing. Overhead, unswept wing.**

Cessna makes more varieties of business jets than anyone else does, and the Encore is its K-car, or soccer-mom minivan, with eight passenger seats and just 4'8" (1.46 m) of headroom.

Cessna Citation I

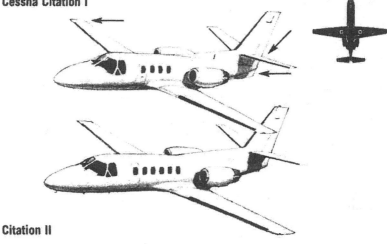

Citation II

Cessna CJ2

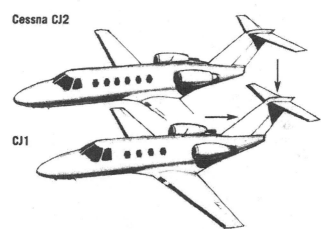

CJ1

Cessna Encore (Ultra)

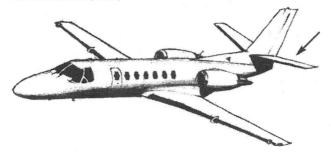

Cessna Citation Excel, XLS
Length: 5'10" (15.79 m) **Wingspan:** 55'8" (16.98 m) **Cruising speed:** 497 mph (800 km/h)

The Excel is substantially larger, especially in fuselage diameter, than the Cessna Encore (previous entry). Its five Cessna-style windows are aft of the passenger door. The substantial wing fairing includes the windowless passenger door. Dihedral in tail planes, two small cockpit side windows.

This Cessna carries the same seven passengers as the other small Cessna with the low-mounted tail plane (Encore, previous entry). The much larger Cessna Sovereign (page 170) seats eight luxuriously. As Cessna continues to make aircraft for all possible customer wish lists, the Excel features more speed and more comfort, with 5'8" (1.73 m) of headroom. It is essentially the same interior (minus one seat) as the T-tailed Citation X (page 170).

Israel Aircraft Industries 1123 Westwind, Commodore, Jet Commander
Length: 52'3" (15.93 m) **Wingspan:** 44'9" (13.65 m) **Cruising speed:** 420 mph (676 km/h)

Fuselage-mounted twin jets; conventional tail; wingtip tanks. (The only other planes with factory tip tanks and twin fuselage-mounted jets are the Learjets, which have T-tails.) High overhead, you can separate these from Learjets by the gap between the wing trailing edge and the engine nacelle (the forward half of the Learjet engines rides up over the wings). The last model, the Westwind II, had winglets on the tip tanks.

A 10-passenger jet designed in 1963. The design was sold to Israel Aircraft after the merger of Rockwell and North American in 1967. Part of the merger agreement required the combined firms to manufacture only one executive jet, and it kept the North American Sabreliner (page 172).

Israel Aircraft Industries 1125 Astra
Length: 52'9" (16 m) **Wingspan:** 52'8" (16.05 m) **Cruising speed:** 500 mph (850 km/h)

This decades-old swept-wing business jet resembles the same-era Falcon 200 (page 176) but the Astra's tailfin rises abruptly from the fuselage, and it has five rectangular windows; the Falcon's are vertical ovals. The lack of a fairing to the tailfin gives the illusion that the tailplane is higher than average, but it isn't. The Falcon's tailplane sits much higher on the fin, and it has a busy six-piece windshield.

When IAI developed this much more conventional-looking jet (compare the Westwind, previous entry) it started down a long path to today's Gulfstream G100, G150, and G200 (page 182) that are still built in Israel and marketed by Gulfstream. In the 1990s, after improvements and some visible changes (winglets), the Astra was renamed the Galaxy, and marketed by a separate coporation also named Galaxy. That company was purchased by Gulfstream, and all subsequent IAI business jets are sold by, and under the name of, Gulfstream.

Cessna Citation Excel, XLS

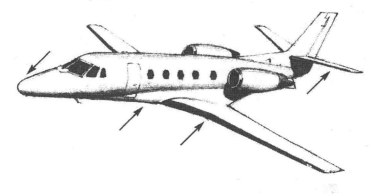

IAI Westwind

IAI Astra

Cessna Citation III, VI, VII, X
Citation III data: Length: 55'5" (16.9 m) **Wingspan:** 53'4" (16.24 m)
Cruising speed: 540 mph (869 km/h)

Separate this series from similar designs by the pleasingly **peculiar sculpted nose that flows into the wing roots, three nontrailing flap guides, "bullet" where the horizontal stabilizer crosses the tail fin, and vertical TV-screen windows.**

A 6- to 10-passenger luxury boss-hauler, certified to fly above the weather at more than 50,000 ft. (15,240 m). The IV is a less luxurious entry-level model; the VII, a higher-performance III; and the X, a longer-range and slightly stretched version, showing six windows behind the portside doorway and seven on the right-hand side of the airplane.

Cessna Citation X
Length: 72'4" (22 m) **Wingspan:** 63'6" (19.4 m) **Cruising speed:** 590 mph (950 km/h)

A very large corporate jet, with seven of the typical Cessna passenger windows on each side: Noticeably swept wings come out of a long aerodynamic fairing that begins in front of and includes the passenger door.

The newest and certainly the largest Cessna with trans-Atlantic range and speed approaching the sound barrier. Only the Gulfstreams (page 182) are larger; although also T-tailed, their horizontal, oval windows are a quick distinguishing mark. It is unlikely that corporate jets, as such, will grow much more. The smaller regional airliners by Embraer and Bombardier (page 188) are already available with plush, not to say fat-cat, interiors, as is the Boeing 717 (their production and new name for the MD90, page 190). If a larger business jet is ever built, it will have extremely high speed to compete with the reconfigured airliners.

Cessna Sovereign
Length: 63'7" (19.38 m) **Wingspan:** 63'2" (19.24 m) **Cruising speed:** 522 mph (840 km/h)

Seven Cessna-style passenger windows each side on a plane almost 10 ft. shorter than the six-windowed Citation X (previous entry) and the **same tail plane configuration as the much smaller Encore and Excels (pages 166 and 168). Slightly swept wings arise from a large aerodynamic fairing** like the Citation X's; the cockpit windows and nose configuration are visually identical in the two.

Technically, the Sovereign is an eight-seater, like the Citation X, but the eighth is a side-facing, non-club-style seat for either the attendant or the most junior executive (who probably will have to serve the drinks if there is no cabin crew). This plane is essentially a slower version of the Citation X, with the same good headroom (5'8", 1.75 m). Slower planes with less-swept wings have one very attractive advantage: They need about three-quarters the amount of runway, at whatever ground level and temperature. A loaded Citation X needs almost 1.5 mi. (2.4 km) on a summer day in Denver, Colorado; the Sovereign, a little less than 1 mi. (1.6 km).

Cessna Citation III

Cessna Citation X

Cessna Sovereign

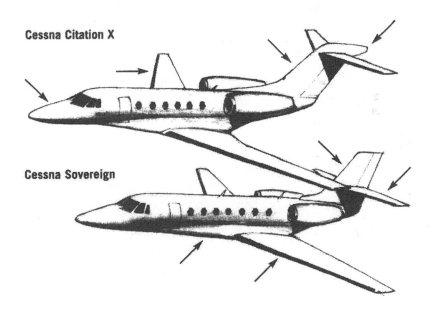

North American Rockwell Sabreliner, CT-39
Model 75 data: Length: 47'2" (14.38 m) **Wingspan:** 44'8" (13.61 m)
Cruising speed: 600 mph (965 km/h)

A series of very similar aircraft with slight dimensional changes. **Twin fuselage-mounted jets; conventional tail; fully swept wings** (the Cessna Citation I and II have straight wings and conventional tail; the Falcon has swept wings with tail planes mounted midway up the fin); **very chubby** compared to similarly sized exec-jets, giving 6 ft. of headroom inside.

Developed in 1958 as a utility and jet trainer for the military (supplied as T-39 and CT-39 to the U.S. Air Force and the Navy), the military Sabreliners and the old model 40 had three triangular windows behind the passenger door. Later stretched versions have five triangular or square windows. The general appearance of the plane remained unchanged by modifications. Accommodates 8 to 12 passengers, depending on seating density.

Lockheed Jetstar, C-140
Length: 60'5" (18.42 m) **Wingspan:** 54'5" (16.6 m) **Cruising speed:** 508 mph (817 km/h)

Uncommon and unmistakable. Combines **four rear-mounted engines** with **massive fuel tanks "glove mounted" on wings.**

Lockheed's partly civil, partly military light transport was produced in small numbers, including 16 Jetstar Is for the U.S. Air Force (they have slightly smaller engines than illustrated. North American's Sabreliner (preceding entry) got most of the military business, and Lockheed stopped building Jetstars in 1981, after 21 years of production. Crew of 2; 10 passengers. Complete airliner appointments, including automatic oxygen mask delivery in case of loss of pressure. A few Jetstars have been converted to twin fan-jets, but fuel tanks are diagnostic.

British Aerospace 125, C-29
Length: 700, 50'8" (15.46 m) **Wingspan:** 47' (14.33 m) **Cruising speed:** 449 mph (722 km/h)

One of two aircraft with fuselage-mounted twin engines and **midway tail plane** (not T-tail); compare with Dassault Falcon 20 (page 176). The 125 has **moderately swept wings** (the Falcon has a strong 30-degree sweep) and **shows a noticeable tail fin fairing rising out of the fuselage over the engines and a ventral fin below the tail** (the Falcon does not). The 125 has five windows, right side; the current model, 700, has six.

A popular business jet; more than 600 of the 125 series sold from 1965 to 1980. Stretched and streamlined model 700 carries as many as 14 passengers. When marketed in the United States by Beech, it was known as the Beech Hawker. The refined model 800 was introduced in 1984.

North American Rockwell Sabreliner

Lockheed Jetstar

BAe 125

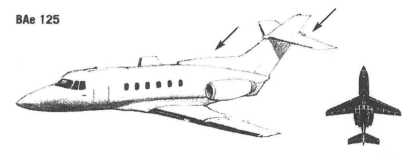

Sino Swearingen SJ30-2

Length: 46'11" (14.3 m) **Wingspan:** 42'4" (12.9 m) **Cruising speed:** 520 mph (840 km/h)

The smallest of the new generation of business jets (about the same dimensions as the original Learjets). Note the **large, gracefully curved tail plane; slim, strongly swept wings; large cockpit windows** compared to the size of the aircraft. **Five vertical, narrowly rectangular passenger windows clustered closely.** What appears to be a large ventral strake is actually a **ventral control fin**; close at hand, the ventral rudder's three pivot points are visible. **Three nontrailing flap guides.**

This clean-sheet-of-paper design has achieved a number of firsts for a very small business jet, including a 20% increase in cabin pressurization (fewer altitude headaches), nonstop catercorner continental range (e.g., Miami to Seattle), and cruise speeds from 12% to 25% faster than similar-sized aircraft. Invisible innovations include fuselage fuel tanks as a noise-dampening factor and a flap system allowing for takeoffs from fields a bit under 4,000 ft. long (1.22 km). A tradeoff has been cabin height: a mere 4'4" (1.31 m).

Hawker Horizon (Raytheon)

Length: 69'2" (21.08 m) **Wingspan:** 61'9" (18.82 m) **Cruising speed:** 485 mph (783 km/h)

The newest Raytheon Aircraft supermidsize corporate jet is characterized by the **largest belly fairing for a plane its size, strongly swept wings,** and a **slight kink in the trailing edges. Seven vertical oval windows on each side; deep windscreen** dominates a relatively **short nose.**

The Horizon is about one-third larger than Raytheon's Hawker 800XP and has the same smoothly curved fuselage and fairings made possible by an all-composite construction. The Horizon will be scarce for some time; first deliveries were in 2004.

Hawker 800XP (Raytheon) Beechjet 800A

Length: 51'2" (15.6 m) **Wingspan:** 51'4" (15.66 m) **Cruising speed:** 515 mph (828 km/h)

One of several business jets with the **tail plane mounted halfway up the tail fin;** compare with its type original, the BAe 125 (page 172) and also the much larger Dassault Falcon 2000 (page 176) and Cessna Sovereign (page 170). All have relatively **low-mounted engines,** but the Falcon's **tail fin rises abruptly from the fuselage,** compared to the long fairing to the tail fin on Hawkers. As time went by, the 800XP and the last 800As ended up with **longer wings, a deeper belly bulge, more oval windows, and a larger, slanted windscreen** than the original 800A.

The persistence of the "European" midway-mounted tail plane and the revival of the Hawker name is about all that shows the ancestry of the 800XP. The long and relatively slim wings make it a "square" plane. Although the fuselage grew by only 4 in., the wings expanded by a little over 4 ft. Marketed as "roomy," it has 5'9" of headroom, not excessive compared to recently designed business jets.

Sino Swearingen SJ30-2

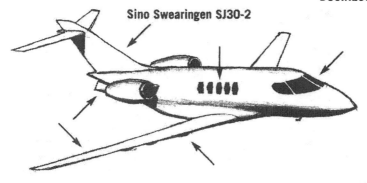

Raytheon Hawker Horizon

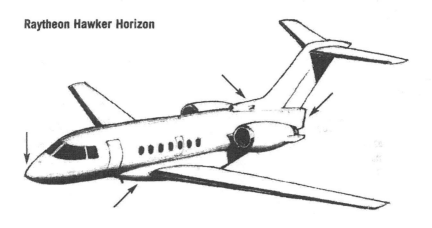

Raytheon Hawker 800XP

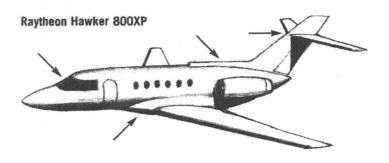

Dassault Falcon 10, 100, 20, 200, HU-25, CC-117
Model 20 data: Length: 56'3" (17.15 m) **Wingspan:** 53'6" (16.29 m)
Cruising speed: 536 mph (862 km/h)

One of two fuselage-mounted twin jets with the **tail plane midway up the tail fin** (compare with the BAe 125, page 172). **Falcon tail fin has a very short fairing; strongly swept wings and tail plane.**

Popular as an executive, airline, and air-cargo plane, the Falcon 20 is being used by the U.S. Coast Guard (HU-25) and Canadian armed forces (CC-117) as a long-range patrol plane. Various passenger and cockpit window configurations, including the solid-bodied cargo craft seen at so many U.S. airports. Model 10s and 100s are 11 ft. shorter in wingspan and length, with either three windows (model 10) or three port and four starboard (model 100). The Falcon 200 is a modified 20 and was introduced in 1984.

Dassault Falcon Late 200, 900, 900EX, 2000, 2000EX
Model 900 data: Length: 63'4" (20.21 m) **Wingspan:** 63'5" (19.33 m)
Cruising speed: 528 mph (850 km/h)

These planes are slightly different from the Falcon 50s and earlier Falcon 900s. **Engine nacelles enclose the exhaust tube** (which was visible in the three-engined 50s, 500s, and earlier 900s). **The new Falcon twin-engine 200 is stretched and shows 10 windows each side. The trailing edge of the new 200's wing shows a distinct break at the midpoint. The intake for the engine on top of the new 900 models is a bit closer to the fuselage and more neatly faired-in.** These new models do keep the **busy seven-piece windshield and the inverted dihedral in the tail plane** of all Falcons still in production.

The newest 900EX version has improved Dassault's standing in the increasingly important long-distance intercontinental market. Although it falls just short of the Paris–Tokyo or London–Los Angeles nonstop of the newest Gulfstreams, it can do Paris–Beijing (which is closer on the polar great-circle route than Tokyo). The 2000EX is trans-Atlantic, as is, with even more reserve, the latest 900, the 900C. The EXs, with subtle engine improvements and aerodynamic tweaks, cannot be visually separated from the 900C or the 2000.

Dassault Falcon 200

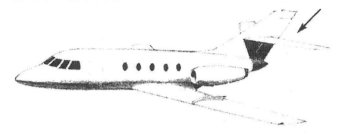

Dassault Falcon 100

Dassault Falcon 900, 900EX, Falcon 2000, 2000EX

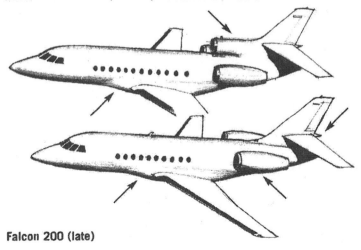

Falcon 200 (late)

177

Dassault Falcon 50, 50EX

EX data: Length: 60'9" (18.5 m) **Wingspan:** 61'10" (18.85 m) **Cruising speed:** 528 mph (850 km/h)

Very minor changes from the basic Falcon 50 design show on the new 50EX (lower sketch): There is a slight break in the leading edge of the cranked wing on the 50EX, and the tail plane is ever so slightly more swept, adding 9 in. (23 cm) to the overall length. Fairings are smoother, and the new engines do not reveal the exhaust tube, as on the 50. In common: **three engines, one mounted through the tail fin like a miniature L1011; tail plane higher up on the fin than on other aircraft** (Citations, BAe125, etc.); **seven windows each side.**

The EX is just a bit faster and bit longer-range than the original 50; at a range of 3,540 mi. (5,695 km), the EX is transcontinental but not really trans-Atlantic. It requires less runway, only 0.9 mi. (1.48 km) at sea level, with a reasonable load of passengers and fuel, than the larger Falcons.

Dassault Falcon 7X

Length: 78'5" (23.9 m) **Wingspan:** 82'5" (25.13 m) **Cruising speed:** 580 mph (934 km/h)

A very large business jet, with **wings noticeably greater in span than the aircraft's overall length.** Like other recent Falcon models, **the tail plane has a distinct droop, a reverse dihedral. The supercritical swept wing has a break in both leading and trailing edges and, like all Falcons, no winglets.** Like a number of the most recent business jets, it's **belly fairing has swollen noticeably** over its Falcon predecessors. It shows **14 windows each side;** the other three-engine Falcons range from 7 to 12 windows each side. Breaking with two decades of practice, the 7X windshield is a simple four-piece, not the typical Falcon seven-piece.

Scheduled for first deliveries in 2005, the 7X's major improvements over the Falcon 900 and the 900EX are a little additional speed and a much greater range: 6,525 mi. (10,500 km), or Paris–Tokyo nonstop.

Dassault Falcon 50

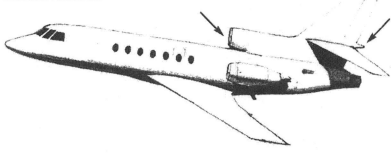

Dassault Falcon 7X

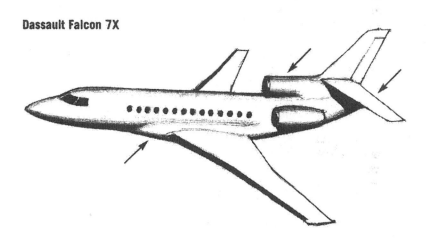

Canadair 600 Challenger, 601, 601-3A, CL601 RJ

Canadair 600 data: Length: 68'5" (20.85 m) **Wingspan:** 61'10" (18.85 m)
Cruising speed: 509 mph (819 km/h)

One of the largest and thickest of the T-tailed business jets. Five visible, deep flap guides on each wing; quite square windows.

A series of jets, all airliner-deep through the fuselage, with headroom for passengers 6 ft. (1.83 m) tall. The 601 has winglets and even larger turbofan engines; the 601-3A has eight windows and a luxury interior for 10 passengers. Newest is the 601 RJ (regional jet), stretched to 88'5" (26.95 m), which carries up to 50 passengers. It shows 13 windows on each side and is the size of the original DC9.

BOMBARDIER BUSINESS JETS: CHALLENGER SERIES

After acquiring Canadair, Bombardier kept the Challenger program going but built one new model, the Challenger 300, from an all-new design. The Bombardier Challenger 604 and its stretched offspring, the Bombardier Challenger 800, are evolved from the original Canadair 600 Challenger. All three are characterized by the original commitment to building a "business airliner" so that the average person can stand upright in the aisle. This was accomplished by placing the wing roots and the landing gear well below the cabin in an almost guppylike swelling, a Bombardier adaptation from the world of jet airliners now more common; see the Cessna Citation X (page 170) as one example.

Bombardier Challenger 300

Length: 66'7" (20.9 m) **Wingspan:** 63'10" (19.5 m) **Cruising speed:** 528 mph (850 km/h)

The shortest Challenger, it appears chubbier than it actually is, a look accentuated by the large, swollen fairing enclosing the landing gear and wing roots. A clean wing, with three flap guides that do not protrude beyond the trailing edge; winglets. Six windows each side; large engines without protruding thrust reversers. The top of the tail fin is bulkier and more protrusive than on the other Challenger aircraft.

Bombardier Challenger 604

Length: 68'5" (20.9 m) **Wingspan:** 64'4" (19.6 m) **Cruising speed:** 528 mph (850 km/h)

A stretched and improved version of the Canadair Challenger, shows one more window each side (seven port, eight starboard rectangular windows), and the cabin interior is a comfortable 8'2" (2.5 m) wide. Typical new Challenger bulging underwing fairing and engines with exposed exhaust nozzle. Clean wing with winglets.

Bombardier Challenger 800

Length: 87'10" (26.8 m) **Wingspan:** 69'7" (21.2 m) **Cruising speed:** 568 mph (914 km/h)

The Challenger 800 is a business-configured Bombardier CRJ200, both characterized by an unsuperstitious 13 windows each side. See page 188.

Canadair Challenger

Bombardier Challenger 300, 604, 800

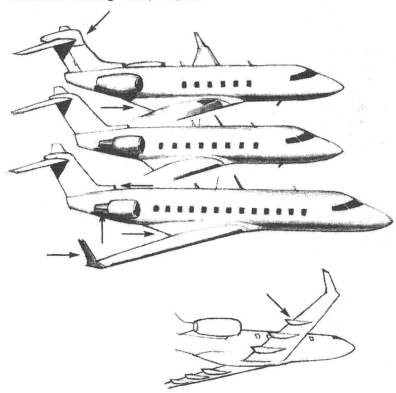

Galaxy Gulfstream, Gulfstream G100, G150, G200
Gulfstream 200 data: Length: 62'3" (18.97 m) **Wingspan:** 58'1" (17.7 m)
Cruising speed: 540 mph (870 km/h)

A long-lived family of mid- to large-sized business jets with some consistency over time: four-piece windshield, round passenger windows, winglets (like almost all swept-wing business jets), three trailing flap guides each side; most tail planes without fairing, but newest have a fairing with a pressure opening. Number of windows varies over time and within model numbers, from five to seven each side. Designed and built by IAI in Israel, the fundamental differences have been increases in cabin height—the 100s have 5'7" (1.7 m); the 200s, a generous 6'2" (1.88 m)—and cabin widths (elbow room) have increased from the original 4'8" (1.42 m) to 7'4" (2.24 m). The fuselage has stretched from 55'7" (16.95 m) to the 200's 62'3" (19 m). Range has increased from 3,100 mi. (4,990km) to 3,914 mi. (6,300 km). The marketing decision to shift from Roman numbers (I, II) also added two zeros.

Gulfstream II, III, IV
Gulfstream III data: Length: 83'1" (25.32 m) **Wingspan:** 77'10" (23.72 m)
Cruising speed: 512 mph (824 km/h)

A huge business jet (two-thirds the size of an unstretched DC9), with shallow oval window, T-tail, and fuselage-mounted twin jets. Accommodates eight passengers and a crew of three. Intercontinental range. Model II did not have winglets. The 1983 introduction, model IV, is 4'6" (1.37 m) longer and shows six rather than five passenger windows.

Grumman designed and built 258 Gulfstream IIs between 1967 and 1969. The U.S. Coast Guard operates one Gulfstream II as a VIP transport, clearly marked with the CG's red diagonal stripe. Don't confuse it with the Coast Guard Falcon 20 search planes, which have the tail plane mounted halfway up the fin and round windows.

Gulfstream G200

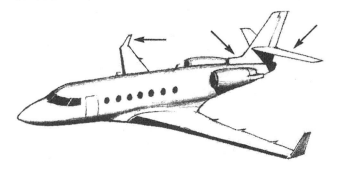

Gulfstream II

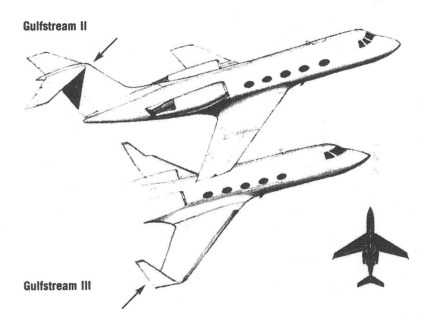

Gulfstream III

Gulfstream GV, G550, G500, GVSP

Length: 96'6" (29.4 m) **Wingspan:** 93'6" 928.5 m) **Cruising speed:** 567 mph (912 km/h)

The longest-range business jet ever, the G550 has seven horizontally oval windows and a swept tail fin that changes angle at the large tail plane. The original model, the GV, had six windows. The wings are noticeably long and thin and have new Gulfstream winglets, which rise halfway back the wing tip. Changes in the GV's engine efficiency and tweaks of the streamlining allow the G550 to fly New York–Tokyo in adverse headwinds, something the GV (the G500) could not do. In perfect conditions, the 550 can fly 7,762 mi. (12,492 km) nonstop.

Gulfstream G450, 350, GIV, etc.

Length: 89'4" (27.23 m) **Wingspan:** 77'10" (23.7 m) **Cruising speed:** 528 mph (850 km/h)

Typical Gulfstream I horizontal oval windows; fairing to tail plane begins halfway along the fuselage; uncomplicated T-tail, miniwinglets.

Gulfstream returned to Arabic numerals in the twenty-first century. There are no visible differences among the aircraft in this series; different fuel capacities and total weights change only the range and the required length of the airfield for takeoff. The 450 has the longest range, at 5,509 mi. (8,061 km). The first models of the Gulfstream V also had six windows each side but much longer wings.

Bombardier Global 5000, Global Express

Global 5000 data: Length: 96'9" (29.5 m) **Wingspan:** 87'5" (26.6 m)
Cruising speed: 562 mph (904 km/h)

A huge business jet, almost the size of the original Boeing 737. The Global Express, at 99'3" (30.2 m), approaches the size of the first stretched Boeing 737. Engines with enclosed exhaust nozzles. Thin, high-performance wings appear longer in proportion to the body than normal, but that is an illusion. Each wing carries three flap guides that trail behind it. At a distance, the trailing edge of the wing appears to be curved; it is actually a series of straight edges. The chief marketing advantage, besides pure size, is the long range: The Global 5000 can cover 5,524 mi. (8,890 km) nonstop. The Global Express can reach 7,077 mi. (11,390 km): that's Chicago to Madrid with fuel to spare.

Gulfstream GV, GVSP

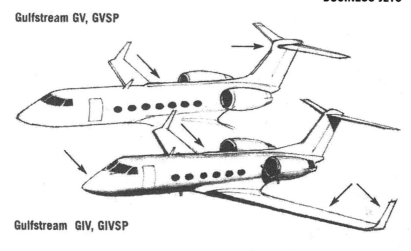

Gulfstream GIV, GIVSP

Bombardier Global 5000, Global Express

185

JET AIRLINERS

Fairchild Dornier 328
Length: 69'10" (21.28 m) **Wingspan:** 68'10" (20.98 m) **Cruising speed:** 380 mph (613 km/h)
Cruising speed for jet: 460 mph (741 km/h)

Except for the engines, two identical airplanes: Dornier's trademark wing, with the leading edge kinked at the engine nacelle; straight trailing edge at a 90-degree angle off the top of the fuselage; T-tail; substantial but not huge belly fairing for the main landing gear.

An outgrowth of the Dornier 228 small twin (page 144), the first 328 was the turboprop; the change to a turbofan jet engine was accomplished in less than a year, as no other structural changes were required. The jet version does take longer to get to takeoff velocity, still a very short 4,000 ft. (6,437 km). The turboprop engines force air over the wing directly, and that model needs a very short 3,500 ft. (5,633 km) of airfield, making it the largest (32-passenger, 3 crew) STOL flying.

BAC 111 (One-Eleven)
Series 500 data: Length: 107' (32.61 m) **Wingspan:** 93'6" (28.5 m)
Cruising speed: 461 mph (742 km/h)

A low-wing, T-tail, fuselage-mounted twin-jet airliner. Note four field marks separating it from the similar DC9 and Fokker Fellowship: combines pointed nose, oval windows, three flap guides that trail behind on each wing, and distinct bullet on tail plane.

Certified in 1965 as a 79-passenger series 200 aircraft, the most common variant in the United States is the stretched series 500, carrying up to 119 passengers. Basically a short-haul aircraft, it is also produced in a variant for small, high-altitude, hot-weather airports: the series 475 — 14 ft. shorter but with the long wings and high power of the stretched 500. Last manufactured under license in Romania.

Fokker F28 Fellowship, 100
Model Mk4000 (Fellowship) data: Length: 97'1" (29.61 m) **Wingspan:** 82'3" (25.07 m)
Cruising speed: 421 mph (677 km/h)

Quite rare in the United States. A stubby, low-wing, T-tail, fuselage-mounted, twin-jet airliner. Separate from the much more common DC9 or BAC 111 by these marks: short, rounded nose; oval windows; distinct fairing from fuselage to tail fin; two flap guides that trail behind on each wing; squared-off rear fuselage housing a clamshell airbrake.

Fokker attempted to carve out a particular market segment with this short-haul, high-performance aircraft. Carrying a maximum of 85 passengers in the Mk4000 configuration, the Fellowship is highly fuel-efficient and suitable for intercity hops of as little as 30 min. flying time. A stretched model 100 carries 97 to 122.

Fairchild Dornier 328Jet, 328

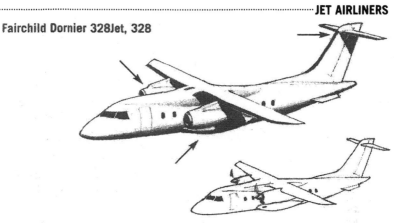

BAC 111 (One-Eleven)

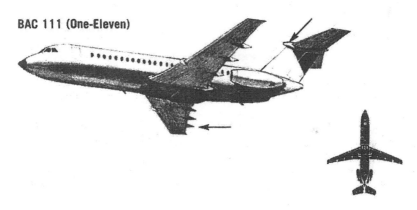

Fokker F28 Fellowship

Embraer 170, 190

170 data: Length: 98' (29.9 m) **Wingspan:** 85'4" (26 m) **Cruising speed:** 540 mph (870 km/h)

A pair of new Embraer regional aircraft the size of a small jetliner, characterized by a **shorter nose than the Embraer regional jets;** quite **square passenger windows; a low-set tail plane, swooping leading edge on the tail fin; noticeable fuselage extension and winglets. A big belly fairing** and turbofan engines with visible exhaust tubes tucked in close to the wing root.

The 70-passenger 170 and the 20 ft. (6.1 m) longer 90-passenger 190 are aimed at a market once controlled by two U.S. aircraft: the original unstretched Boeing 737 and the McDonnell Douglas DC9/MD10. The company plans extended-range versions, but the standard 170 with a full load has a range of 2,400 mi. (3,862 km), which would get it from London to Moscow or from Dallas to Halifax, Nova Scotia. It has a remarkably short takeoff run of just 4,000 ft. and reasonably quiet engines; the combination of agility and good manners makes it possible to use smaller airports, such as London City or Florence.

Embraer ERJ 135, 140, 145

ERJ 145 data: Length: 98' (29.87 m) **Wingspan:** 65'7" (20.04 m) **Cruising speed:** 518 mph (833 km/h)

A distinctive design: **Swept wing is strongly kinked on the trailing surface; substantial belly fairing and a rather thick fairing at the wing root, no winglets. Three flap guides each side barely extend beyond the trailing edge.** It looks much more stubby winged than it actually is, although its wings are some 5 ft. shorter each side than a comparable-sized Bombardier CJR; their width and thick fairing at the wing root increase the sense of shortness. Especially parked facing the waiting lounge at the airport, **all Embraer ERJs from 135 to 145 have noticeably long, pointy noses.**

The three members of this group of Brazilian-designed and -built airplanes carry, respectively, 37, 44, and 50 passengers. Except for the extended-range models with additional fuel aboard, the ERJs of this type need considerably less than a mile of runway for takeoff. The 135 is available as a luxury outfitted business jet.

Bombardier CRJ 100, 200, 700, 900

200 data: Length: 87'10" (26.77 m) **Wingspan:** 69'7" (21.2 m)
Cruising speed: 488 mph (786 km/h) **900 data: Length:** 119'4" (36.37 m)
Wingspan: 76'3" (23.24 m) **Cruising speed:** 528 mph (850 km/h)

The C(anadair) R(egional) J(et) is essentially a stretched business jet, the Canadair 600 Challenger. An unusual combination for a regional jetliner: slim wings with prominent winglets, a T-tail, fuselage-mounted jet engines with exposed exhaust tubes. The 900 appears to have done most of its stretching in front of the wings.

When it acquired Canadair, Bombadier's first model was essentially the Canadair Challenger 601, renamed CRJ 100. The enormous stretch came in stages, when the model 700 appeared at 105 ft.

Embraer 170

Embraer ERJ 145, ERJ 135

Embraer ERJ 135

Bombardier CRJ 900

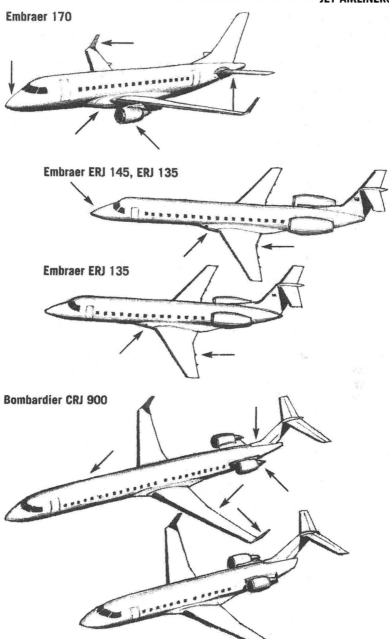

Bombardier CRJ 200

McDonnell Douglas DC9, MD80 to MD90
MD80 data: Length: 147'10" (45.06 m) **Wingspan:** 107'10" (32.87 m)
Cruising speed: 565 mph (909 km/h)

If you see a **medium-length to very long airliner with two rear-mounted engines**, it will, 99 times out of a 100, be a version of the DC9, MD80 to MD90 aircraft, now the Boeing B717. They range from old 50-passenger models to stretched MD90s that carry 158 passengers. To separate out the ones that are similar in size to the BAC 111, note the **absence of a bullet where the tail plane crosses the tail fin**, and note the **flap guides that do** *not* extend past the wing's trailing edge. The rare Fokker Fellowship (page 186) has a much stubbier nose, two trailing flap guides, and a curious squared-off tail. Compare the new regional airliners, page 188.

Boeing 717
Length: 124' (37.8 m) **Wingspan:** 93'3" (28.4) **Cruising speed:** 504 mph (811 km/h)

Some subtle changes from its ancestors, the DC9s and the MD80s and MD90s: At the terminal, you'll see that **the 717 has added a pair of small look-up windows for the pilot and copilot.** If there's a DC9 or an MD in the area for comparison, the **717 has considerably larger engines.** The leading edge of the tail fin curves up more abruptly; at the tail plane, the leading edge is almost vertical to the fuselage. The fuselage extension below the tail fin is squared off, as were the extensions on the entire MD series. Only the DC9 had the curved end to the fuselage extension.

The most prolific of airliners, at least in models, the Boeing 717 was the last version in production. It is something of a retro: shorter than any of the MD series, shorter than the largest stretch of the older DC9. This is not the first time that the manufacturer, once McDonnell Douglas, then Boeing (which acquired McDonnell Douglas), has opted for a modestly stretched airplane.

Boeing 737-200, -300, -400, -500
737-300 data: Length: 109'7" (33.4 m) **Wingspan:** 94'9" (28.88 m)
Cruising speed: 564 mph (907 km/h)

Even when stretched a little (models 300, 400, 500), a **stubby twin underwing jet** that can, from a distance, appear to be a wide-body. The original 737-200 has **slim engine nacelles that extend equally in front of and behind the wing.** The 300, 400, and 500 have large-diameter fan-jets mounted on pylons. Overhead, where relative size is difficult to judge, they could be confused with Airbus A300s or 767s. The 737 has three flap guides; the 767 has four almost invisible guides; the A300 has five noticeable guides.

The primary short-haul jet of the 1970s, it carried only 120 passengers then, and even the stretched (110 ft., 33.53 m) 400 carries only 146 today. It has excellent short-field qualities, and a number of 200s were modified for use on gravel airstrips in Africa, the Middle East, and Latin America.

McDonnell Douglas DC9

Boeing 717

Boeing 737-300

Boeing 737-200

191

Boeing 757

Length: 154'8" (47.14 m) **Wingspan:** 124'6" (37.95 m) **Cruising speed:** 494 mph (795 km/h)

Slim-bodied, with two large turbofans mounted under the wing, showing well forward of the wing. This plane should separate easily from the wide-bodied, twin-turbofan airplanes, but compare with the Airbus A-300 and the Boeing 737 and 767. The combination of normal fuselage and engines is diagnostic.

From the passenger's point of view, the 757 is nothing more than a stretched, re-engined version of Boeing's popular 727 aircraft. Other differences are subtle but include a wing with less sweepback and greater depth (chord). The 757 is 19 ft. longer than the 727. Like almost all new airliners, the 757 carries more passengers and is certified to fly with two rather than three flight officers — a considerable saving.

Airbus A320

Length: 123'3" (37.57 m) **Wingspan:** 111'3" (33.91 m) **Cruising speed:** 515 mph (829 km/h)

Not much bigger than a 737 but a truly wide-bodied medium-length twin jet. Like all the Airbuses, it has the very noticeable flap guides. The winglets are wing fences: up-and-down winglets. Noticeable double-bubble cross-section at wing root for baggage and containers. Like other Airbuses, the top of the fuselage runs into the tail on a nearly straight line, giving the appearance of greater upsweep on the bottom of the fuselage. Boeings are tapered symmetrically, like ice cream cones.

Conventional enough on the outside, the Airbus A320 is highly advanced internally, with the first fly-by-wire controls in the subsonic industry, operated with side-stick controllers (no control columns or wheels in the cockpit). It has wide aisles for easy passenger exiting (and thus quick turnarounds) and has wider and deeper seats than on older airplanes of any manufacture.

Airbus A300, A310

A300-600 data: Length: 177'5" (54.08 m) **Wingspan:** 147'1" (44.84 m)
Cruising speed: 543 mph (875 km/h)

A pair of wide-bodied, twin underwing-engined airliners (the A310 is shorter by 24'4", 7.42 m), separated from their Boeing look-alikes by several marks: Airbuses have large flap guides that trail behind the wing, no dihedral in tail plane, and a typical rear fuselage that extends straight back on the top, with all the taper taken up by the bottom, giving the illusion of an upturn in the tail section. Latest models have little triangular "wing fences" (up-and-down winglets).

These remarkably similar aircraft are available in several subspecies, including longer-range (extra fuel tanks) and convertible, cargo-to-passenger configurations.

Boeing 757

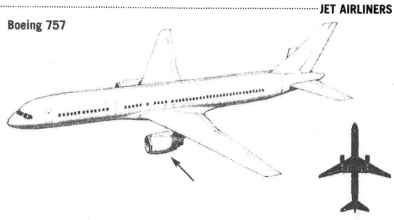

Airbus A320

Airbus A300

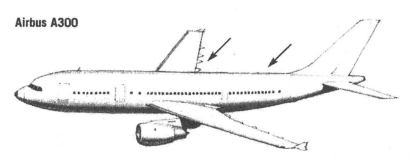

Boeing 767-200, -300
767-200 data: Length: 159'2" (48.51 m) **Wingspan:** 156'1" (47.57 m)
Cruising speed: 494 mph (795 km/h)

A pair of twin wing-mounted jet airliners (the 767-300 stretches to 180'3", 54.94 m), not too difficult to distinguish from the Airbuses: **Three barely noticeable flap guides** (not five obvious ones); **tail of fuselage tapers symmetrically beneath the tail fin; noticeable dihedral in tail planes.** A subtle difference but clear when planes are on the ground: **Where the trailing edge of the Boeing tail fin meets the fuselage, it is forward of the trailing control surfaces on the tail planes.**

The more than 20 models of 767s, including the obvious stretches and the ones with extra internal fuel tanks, can carry from 240 to 300 passengers and operate at ranges from 3,708 mi. (5,967 km) to 7,836 mi. (12,611 km). That kind of doubling of performance, with gradations along the way, creates a plane for every airline's needs.

Boeing 777
Length: 209'1" (63.73 m) **Wingspan:** 199'11" (960.93 m)
Estimated cruising speed: 494 mph (795 km/h)

Began flying in 1994, the largest of the twin-engine jumbos. On the ground, it can be distinguished from the 767s by sheer size and by the two passenger doors in front of the wing (767-300s have an escape door over the wing). **Bulky fairings at the wing roots somewhat resemble Airbus configuration.**

The 777 fills the gap between the stretch 767-300 and the massive 747-400 in both number of passengers and range. The long-range 777 can take 313 passengers in the typical three classes (the 747-400 hauls 386 in three classes) and fly them from London to Los Angeles nonstop.

Boeing 787
Preliminary data: Length: 182' (56 m) **Wingspan:** 193' (59 m) **Cruising speed:** 560 mph (902 km/h)

When built, the 787 will be remarkable for its all-composite curvaceous surfaces and leading edges. Swept wings continue into curvy winglets; all the leading and trailing edges of wings and tails are curves. The nose seems short, an optical illusion from the long, sloping curve of the fuselage from the windshield back to the maximum diameter of the fuselage. Expect it to sport two very large engines.

This very large airplane, approaching the length of Boeing's longest plane, the 777, will carry only 200 passengers in a three-class configuration, the same as an unstretched 757, compared to 360 passengers in the 777. Boeing expects the 787 to create a market for medium-sized but very long-range and fuel-efficient airliners. Boeing predicts that the 787 will use 20% less fuel than a comparably sized aircraft. Interior details are sketchy, but with its wide 18 ft. cabin and considerable length, the 200 passengers should not be as crowded as is common.

Boeing 767-200, -300

Boeing 777

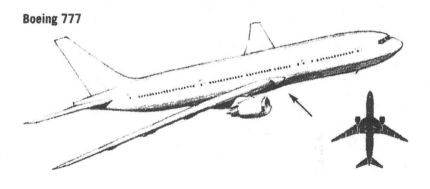

Boeing 787

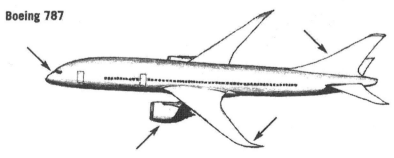

195

Boeing 727

Length: 153'2" (46.69 m) **Wingspan:** 108' (32.92 m) **Cruising speed:** 570 mph (917 km/h)

In North America, the only airliner you'll see with **three rear-mounted engines: one in the tail and the others on fuselage pods.** If someone should import a British Trident, it will have a distinct bullet at the center of the tail plane. The Russian military TU154 should not appear at all, but if seen elsewhere, note that it has a long, pointed bullet at the tail plane.

First flown in 1963, the 727-100 (length, 133'2", 40.58 m) sold moderately to U.S. customers for medium-range flights. Since the introduction of the 727-200, which is 20 ft. longer than the 727-100, Boeing has sold nearly 2,000. As many as 189 passengers can fit, without much comfort, into a one-class 727-200, 90 more than in the original 727-100.

McDonnell Douglas MD10 (DC10), KC-10 "Extender," MD11

MD10 data: Length: 182'1" (55.5 m) **Wingspan:** 165'4" (50.41 m)
Cruising speed: 540 mph (869 km/h)

Common at all large airports. A **wide-body with two wing-mounted engines and a tail engine that blows straight through the tail fin, above the fuselage.** In military paint, it's an Air Force in-flight refueling plane.

First of the tri-engines to carry passengers (1971) and built in a variety of performance models, mostly by changing engines rather than general configuration. A newer MD11 is 18 ft. (5.5 km) longer (200 ft., 60.96 m, overall) and has 323 seats in the usual three classes.

Lockheed L1011 TriStar

Length: 177'8" (54.17 m); Model 500, 165'8" (50.5 m) **Wingspan:** 155'4" (47.35 m)
Cruising speed: 558 mph (898 km/h)

A jumbo wide-body: two engines on wings and one rear-mounted at tail. Separate from the DC10 by noting that the **tail-mounted engine has intake above the fuselage and exhausts through end of the fuselage.** Compare with the DC10 tail engine, which carries straight through the tail fin.

A popular wide-body that has never suffered from a single serious mechanical defect, the L1011 was sold for only 10 years (1972–1982) before Lockheed withdrew from the passenger jet field, leaving it to Boeing and McDonnell Douglas, whose DC10 was a direct competitor to the L1011. Fewer than 300 are in service. The long-range model 500 is not noticeably shorter, but it can be picked out on the flight line by the way the tail engine is faired directly into the fuselage (see sketch above model drawing).

Boeing 727

McDonnell Douglas MD10

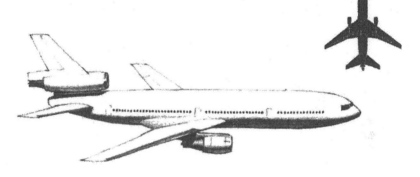

Lockheed L1011 TriStar

British Aerospace BAe146-100, 200, 300, Avro RJ
Length: 93'8" (28.55 m) **Wingspan:** 86'5" (26.34 m) **Cruising speed:** 440 mph (708 km/h)

Smallest of the four-engined jets; **massive fin to T-tail** (not unlike de Havilland's Dash turboprops); **large flap tracks underwing — bulging landing gear fairings on belly;** the only four-jet-on-the-wing T-tail.

Designed over several years, beginning in 1973, by the ailing British aerospace industry, the BAe146 is a short-haul jet that takes advantage of modern fan-jet engines to produce a quiet aircraft; it can land and take off in cities without annoying airport neighbors. First U.S. purchase by Air Wisconsin. Production continued fitfully until ceasing in 2002. They are a fairly common sight at airline hubs in the Midwest and Far West but are now greatly outnumbered by the newer twin-engine regional jets.

McDonnell Douglas DC8
Series 60 data: Length: 187'5" (57.12 m) **Wingspan:** 148'5" (45.23 m)
Cruising speed: 600 mph (965 km/h)

A series of **rare four-engine jetliners.** Compare with the Boeing 707, 720 (next entry) before deciding. The most common variant is the extreme stretch Series 60: Viewed at any distance, it has the aspect of great fuselage length balanced on relatively negligible wings. On the ground, the tail fin has no vhf radio antenna (compare with the 707 drawing); **smooth, cigarshape engine nacelles; distinct "brow" at cockpit window; tail fin swept; but stretch 60 series even more radically swept.**

A popular airliner first flown in 1958. Whatever their original size, most have been converted into the superstretches by the insertion of fuselage plugs fore and aft of the wings. Series 70 is a stretch with more efficient, quieter fan engines. Still flown, mostly as economy charters.

Boeing 707, 720
707-320 data: Length: 152'11" (46.61 m) **Wingspan:** 145'9" (44.42 m)
Cruising speed: 550 mph (885 km/h)

The very rare 707 has a superficial resemblance to the Douglas DC8, but once you have identified the plane by some minor details, its configuration is quite different and instantly recognizable: **four engines on wing and engine nacelles that are distinctly larger forward** (compare with the much smoother, cigarlike DC8 nacelles). **The engines are tucked up under the wing** (the DC8 engines are carried a bit lower and a bit farther forward). **The cockpit windows are very close to the nose** (the DC8 has more nose to it).

The first U.S.-built jetliner, flown in 1954. Very successful; made in a number of variations for increased passenger capacity or for transoceanic flights. The similar 720 was a medium-range plane with thinner, slightly more swept wings and a distinct ventral fin. Alas, a few 707s (model 420) also carry the ventral fin. To muddle the issue further, American Airlines designated its 720s as 707-023s. The airframe is still built for military use as a long-distance radar platform and communication snooper and suppressor.

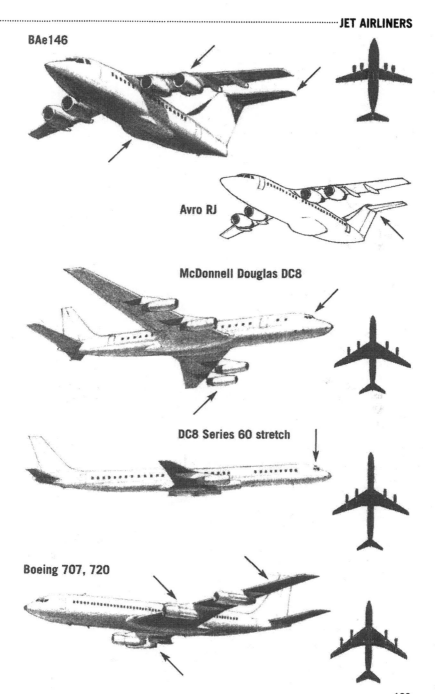

BAe146

Avro RJ

McDonnell Douglas DC8

DC8 Series 60 stretch

Boeing 707, 720

Boeing 747-200, 747SP, 747-300, 747-400
747-200 data: Length: 231'10" (70.66 m) **Wingspan:** 195'8" (59.64 m)
Cruising speed: 604 mph (973 km/h)

Common and unmistakable: the **four-engine jumbo jet with the bulge behind the cockpit** (center drawing). Variants are rarer and more challenging to identify: The stubby, long-range SP (47 ft., 14.33 m, shorter; lower sketch), the stretched upper decks of the 300 (top sketch) and new 400 accommodate more passengers upstairs; and the 400 has winglets.

In 1992, the 747 replaced the old 707 as **Air Force One,** designated VC-25A. Additional 747s purchased by the Air Force from Pan American were converted to long-range cargo haulers, C-19A, with a side cargo door.

Airbus A340, A330
A340 data: Length: 194'10" (59.39 m) **Wingspan:** 192'5" (58.65 m)
Cruising speed: 605 mph (974 km/h)

The A340 is easy to identify: It is the **giant four-engine jet** that doesn't bulge behind the cockpit. Unfortunately, the A330 has an identical airframe, with only two engines, right where the in-board engines are on an A340. Compare the A340 to the similar Boeing 777 (page 194): The A330 has **swept-back and out-canted winglets.** Like other Airbus designs, **the top of the fuselage carries a straight line from the cockpit to the tail fin;** compare to the conical Boeings.

Engineers took advantage of the Airbus configuration to build this competitor to the Boeing 747. The fuselage sections are nearly identical, as are the fly-by-wire controls. The 335-passenger A340-300 can compete over the Atlantic and throughout Europe; however, it has a considerably shorter range than the 747-300: 5,300 mi. compared to the 747's 7,020 mi. (8,525 km to 11,297 km).

Airbus A380
Length: 239'3" (73 m) **Wingspan:** 261'10" (79.8 m) **Cruising speed:** 542 mph (870 km/h)

Enormous. Its tail rises to the height of an eight-story building (79'7", 24.1 m), and the tail itself is 48 ft. (14.6 m) tall. The tail is longer and has much more surface area than a wing on the Airbus Beluga (page 202). A new wing for an airbus, with **no visible flap guides. Four giant fan-jet engines. A number of doors on both the upper and lower cabins,** all out of reach of existing airport equipment. Substantial belly fairing at the wing roots.

Capable of carrying from 550 (three class) to more than 600 passengers, the 380 will require everything from new terminal passageways to thicker concrete runways. Still, Airbus has hundreds of firm orders and expects to deliver the A380 in 2006. The visible cylindrical fuselage conceals two separate pressure-vessel tubes. The out-of-sequence jump to number 380 was intended to reflect a cross-section view of the paired interior fuselages.

200

Boeing 747-300

Boeing 747

Boeing 747SP

Airbus A340

Airbus A380

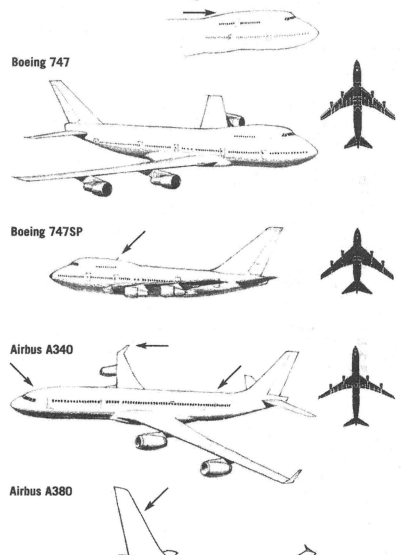

Airbus A300-600ST "Beluga"
Length: 184'3" (56.15 m) **Wingspan:** 147'2" (44.84 m) **Cruising speed:** 460 mph (744 km/h)

Although no identification marks are really necessary, the Beluga carries the typical Airbus wing, with trailing flap guides and an almost indiscernible up-and-down winglet fence. The tail plane is retro in appearance; the only other aircraft with both a large vertical tail fin and outboard fins on the tail plane is the tiny, extinct Grumman OV-10 Mohawk (page 230).

Nicknamed immediately by Airbus engineers after the Beluga whale, five of these enormous carriers are in worldwide charter service, ferrying everything from tanks to fully assembled helicopters. The dropped cockpit allows for a single enormous cargo door at the front of the fuselage, where something as large as 16 ft. by 16 ft. in cross-section (4.88 m by 4.88 m) can be stuffed in the plane. The Beluga replaces the Airbus consortium's 50-year-old Boeing Stratocruiser in moving parts around Europe—it was the last Stratocruiser (based on the B-29) in service.

Aerospatiale/BAC Concorde
Length: 203'9" (62.1 m) **Wingspan:** 83'10" (25.55 m) **Cruising speed:** 1,336 mph (2,150 km/h)

Out of service: long, skinny fuselage with delta wings; four rectangular air intakes under wing; no tail planes at all.

First flown in 1971; first passengers, 1975. After environmental complaints about sonic booms and upper-atmosphere air pollution, airport noise, and the quadrupling of the price of petroleum, the once-hopeful supersonic Concorde was dropped by every airline (more than 70 were on order at one time), except for the government-subsidized airlines of the manufacturing countries, British Airways and Air France. Carried 128 passengers across the Atlantic in less than 3 hours.

Airbus A300-600ST "Beluga"

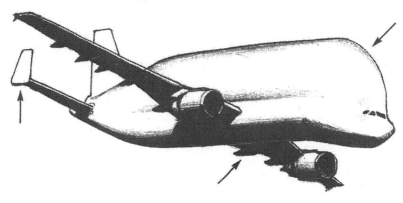

Aerospatiale/BAC Concorde

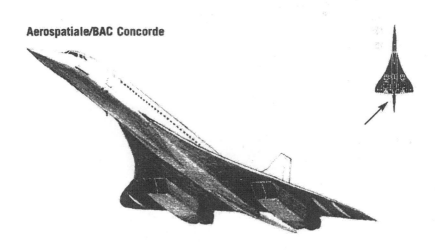

MILITARY AIRCRAFT

Beech T-34C Mentor
Length: 28'8" (8.72 m) **Wingspan:** 33'4" (10.16 m) **Level flight:** 241 mph (388 km/h)

The Navy's only slim-nosed, propeller-driven airplane. High greenhouse canopy; ventral fin; finlet fairings to tail plane; paired air scoops; large side exhausts.

`The latest in a long line of Navy-style in-line trainers, including the SN-J (Texan) and the nonturbocharged Beech T-34 it replaces (page 54). The T-34C, with turboprop, is 90 mph faster than the T-34, making it an easier step up to the 343 mph T-6A used for carrier training (next entry). As with many trainers, it can be fitted with armaments and sold overseas for counterinsurgency missions.

Raytheon T-6A Texan II (U.S.), CT-156 Harvard II (Canada)
Length: 33'3" (10.13 m) **Wingspan:** 32'2" (9.8 m) **Cruising speed:** 345 mph (555 km/h)

This interesting design creates a plane that looks shorter than it really is, owing to the long, high bubble canopy set in the center of the fuselage. Features a slight but discernable gullwing, large side-facing turboprop exhausts; fairing to tail fin, beginning at the canopy; a pair of strakes under the fuselage and a thin triangular fairing to the tail plane.

Based by Beechcraft (now Raytheon) on a Swiss military trainer, the Pilatus PC-9. The U.S. and Canadian version required extensive but invisible improvements, including a pressurized cockpit, ejection seats, improved bird-strike resistance, and propeller deicing system. It is the second-level (after primary) flight trainer in the United States and Canada and is fully aerobatic, including up to a minute of continuous inverted flying.

Pilatus PC-7
Length: 33'4" (10.13 m) **Wingspan:** 33'6" (10.2 m) **Cruising speed:** 289 mph (465 km/h)

A slower brother of the Pilatus PC-9, the PC-7 resembles the Raytheon T-6A (previous entry), including the gullwing and the thin triangular fairing to the tailfin. A rather small bubble-canopied cockpit, a short, thin fairing to the tailfin, and unusual flap on the tail plane extend past the fixed portion.

The PC-7 has been an extremely popular basic (second level) trainer in a couple of dozen countries. In Mexico it was purchased as a trainer but several have been converted in-country to counterinsurgency armed planes.

Beech T-34C Mentor

Raytheon (Beechcraft) T-6 Texan II

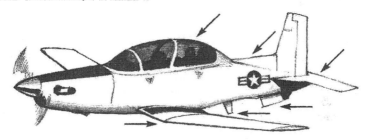

Pilatus PC-7

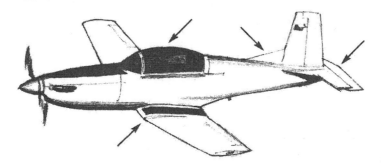

Grumman E-2 Hawkeye and C-2 Greyhound

Length: 57'7" (17.6 m) **Wingspan:** 80'7" (24.6 m) **Cruising speed:** 296 mph (476 km/h)

The E-2 is an unmistakable **twin-engine** aircraft backpacking a **30-ft.-diameter radar pancake.** The C-2 utility version is the **only high-wing twin prop with four tail fins.**

The Hawkeye's mission is early warning for the carrier fleet. The Greyhound serves as a shore-to-ship delivery system, carrying up to 39 passengers or 4 tn. freight. The type has certain Grumman characteristics, including a dihedral in the tail planes and engines that angle out slightly from the fuselage. (Note those features in Grumman's smaller OV-10, page 230, which has three tail fins.) Overhead, it is the only twin-engine propeller aircraft that combines a straight trailing edge to the tail plane with symmetrically tapering wings.

de Havilland CC-115 Buffalo

Length: 79' (24.08 m) **Wingspan:** 96' (29.26 m) **Cruising speed:** 208 mph (335 km/h)

Extremely rare; look for it in British Columbia with Royal Canadian Air Force insignia. **Combination of twin turboprop engines, upswept fuselage, and T-tail is unique.**

The last of two great short takeoff and landing craft (the other was the much smaller Caribou) built by de Havilland of Canada; still serves as a search-and-rescue wing in British Columbia, where its ability to fly in almost any weather has made it unreplaceable, to date. No longer in U.S. service. On the odd chance that you see a civilian aircraft that resembles a small Buffalo but has a kink in the fuselage where it begins to sweep upward, and a midmounted tail plane, you've seen the elusive Caribou.

Grumman E-2 Hawkeye

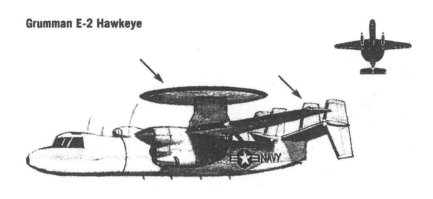

de Havilland CC-115 Buffalo

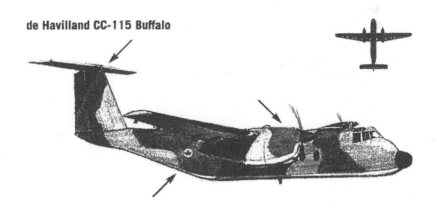

Lockheed C-130 Hercules

Length: 97'10" (29.78 m) **Wingspan:** 132'7" (40.41 m) **Cruising speed:** 340 mph (547 km/h)

Common, nationwide. Combines upswept fuselage with an enormous conventional tail, radar dome nose, and classic Lockheed wing (straight leading edge at right angles to fuselage); four turboprop engines.

The bulky C-130 bears no real resemblance, even overhead, to the more elegant and T-tailed de Havilland Dash 7 (page 154). (There is a Russian copy of the Hercules, the An-12 Cub.) Compare the overhead view of the Hercules with the Electra (page 158). The Hercules is bulkier, and its radar dome nose looks comical. The Orion's (next entry) is simply the curved nose of the airplane. C-130s are operated by all four U.S. services in modes from gunships to weather observation and search and rescue, as well as transports. The Israeli government used the C-130 on the successful mission to free the hijacked Air France passengers at Entebbe, Uganda, on July 3, 1976. Several C-130s are used in Canada for search-and-rescue missions.

Lockheed P-3 Orion

Length: 116'10" (35.61 m) **Wingspan:** 99'8" (30.37 m) **Cruising speed:** 378 mph (608 km/h)

Unique but variable aircraft, seen worldwide. Four long-nacelled propeller engines project well forward of the wings; engines are set well out on relatively short wings; most models show the trailing magnetic detection boom used in antisubmarine patrols. A few carry a round pancake radar dome above the wing; a few show neither magnetic boom nor radar dome, including NOAA weather planes based in Florida.

This is the old Lockheed Electra airliner, first converted to military use in 1959 and still in production after 42 years, flown by dozens of countries for coastal surveillance. Strong and durable, it routinely flies through hurricanes to gather weather data. A proposed successor, the P-7A, will look virtually identical to the P-3; although slightly longer (by 6'4", 1.93 m) and with considerable hidden changes in construction and weapons bays, it would clearly be an upgraded P-3 if it is built.

Lockheed C-130 Hercules

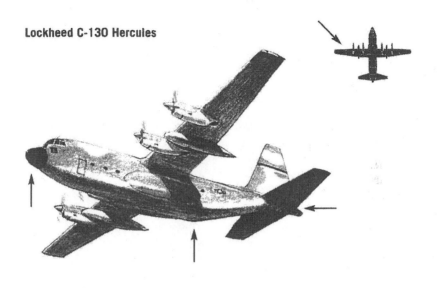

Lockheed P-3 Orion

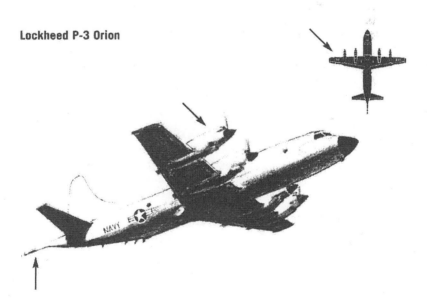

Cessna A-37 Dragonfly and T-37

Length: 29'4" (8.92 m) **Wingspan:** 33'7" (10.3 m)
Level flight: 507 mph (816 km/h) Mach 0.658 at sea level

Low straight wings with conspicuous tip tanks and inconspicuous twin jets at the wing roots; bulbous cockpit for side-by-side seating in the trainer version. Nothing else flying has **twin wing-root jets and straight wings at right angles to the fuselage.**

Although many combat aircraft have been converted to trainers, the counterinsurgency A-37B was developed as a gunship from the U.S. Air Force's standard jet trainer, the T-37. It saw wide use in areas of Vietnam not defended by surface-to-air missiles, carrying a 7.62 mm minigun capable of firing 6,000 rounds a minute, as well as cluster and phosphorus bombs. Suitable for use against lightly armed "insurgents," the A-37's low stall speed, under 100 mph (161 km/h), makes it a precision instrument.

Canadair CT-114 Tutor

Length: 32' (9.75 m) **Wingspan:** 36'6" (11.3 m) **Cruising speed:** 474 mph (763 km/h)

An extremely clever design from Canadair (now subsumed by Bombadier) as a safe, stable, subsonic trainer for the Canadian Air Force. The short **T-tail and the large bubble canopy, along with air intakes** that are large for a single turbojet engine, are a unique combination (the Cessna Dragonfly, preceding entry, has **two engines**).

Flown only in Canada, except for 20 armed versions sold to the Malaysian Air Force, it is still seen at North American airshows, flown by the Canadian Air Force Snowbirds, a nine-airplane close-formation act. In addition to the Snowbird base at Moose Jaw, Saskatchewan, a few are in use as test beds at the Canadian Department of Defense aerospace engineering test establishment at Cold Lake, Alberta. Visitors to Cold Lake might see an even rarer plane, one of the four remaining CT-133 Silver Stars in government use. It is a tandem-seat stretch of the U.S. F-80 Shooting Star, the Air Force's mainstay during the Korean War.

Rockwell T-2 Buckeye

Length: 38'4" (11.66 m) **Wingspan:** 38'10" (11.62 m)
Level flight: 522 mph (840 km/h) Mach 0.69 at sea level

Seen near naval flight schools and stateside aircraft carriers. **Large canopy for tandem pilot and instructor; straight wings with tip tanks; a stubby, front-heavy look.**

The Navy's basic jet trainer used for teaching pilots to land on an aircraft carrier. It resembles the side-by-side-seating U.S. Air Force T-37 if the wing geometry is not visible. The T-2's engine intakes are well forward of the wing. First built as a single-engine trainer by North American, based on the Navy's retired FJ-1 Fury fighters. The twin version is all that flies today, and later models are the first Navy planes with fiberglass wings. Rockwell also markets it as a counterinsurgency plane.

210

Cessna A-37 Dragonfly

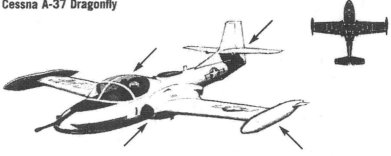

Canadair CT-114 Tutor

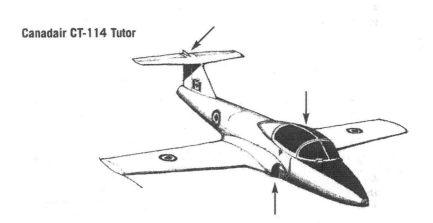

Rockwell T-2 Buckeye

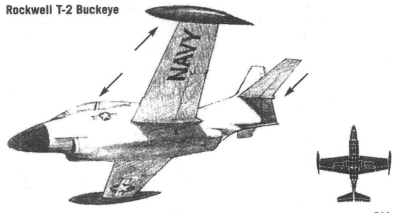

Fairchild A-10 Thunderbolt, "Warthog"
Length: 53'4" (16.25 m) **Wingspan:** 57'6" (17.53 m)
Level flight: 443 mph (713 km/h) Mach 0.58 at sea level

Huge fuselage-mounted turbofan twin jets rise above the fuselage; overhead, note the **rectangular tail plane.**

The A-10 is a highly maneuverable ground-support plane, essentially an aircraft wrapped around a 30 mm gun that fills the inside of the fuselage. The ammunition is typically simple depleted (not radioactive) uranium cylinders that destroy tanks by mere impact. The A-10, basically an alternative to smart bombs and heat-seeking missile systems, relies heavily on the pilot instead of on sophisticated instrumentation for success. Used widely in the 1991 Gulf War against armor and Scud missile launchers and in the second Gulf War as an anti-insurgency tactical attacker.

BAe T-45 Goshawk
Length: 36'8" (11.17 m) **Wingspan:** 30'10" (9.39 m) **Level flight:** Mach 0.85

A distinctive small jet seen near naval bases and aircraft carrier ports: **Long, low bubble canopy covers tandem seating; noticeable reversed dihedral in tail planes.** Up close, the carrier-landing modifications include a tail hook, ventral fin, and leading-edge slats on the wings.

The original BAe Hawk was a British airstrip-based jet trainer, now much modified as a U.S. Navy aircraft carrier trainer. The major visible external change was the addition of leading-edge slats for better stall speeds and quicker rebounds from touch-and-go exercises (including inadvertent touch-and-go "bolter" landings). With numerous modifications, it is the first originally land-based aircraft to be successfully converted to the complex task of flying on and off a moving carrier deck.

Grumman A-6 Intruder/EA-6 Prowler
A-6 data: Length: 54'7" (16.64 m) **Wingspan:** 53' (16.15 m)
Level flight: 625 mph (1,006 km/h) Mach 0.82 at sea level

The **twin jet engines mounted at the wing roots,** combined with swept wings, are diagnostic and give the plane its characteristic look: **bulky forward, slim aft.** Up close, note the hooked-nose electronic probe in front of the cockpit.

The Navy's basic night/all-weather bomber since 1960, the A-6A was heavily used during the Vietnam War, along with the newer Air Force F-111s for night precision bombing. The basic airplane, with side-by-side seating, has been modified into a radar and communications-jamming craft, the EA-6. A four-seat version, the EA-6B, has sophisticated antielectronics capacity. Both E versions are distinguished by the electronic pod on the tail fin; what appear to be externally mounted bombs on the EA-6s are additional wing-mounted electronics. Electronic-jamming Prowlers served in both Gulf Wars.

Fairchild A-10 Thunderbolt

BAe T-45 Goshawk

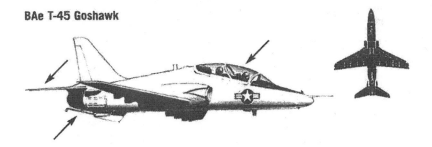

Grumman A-6 Intruder

Intruder

Grumman EA-6B Prowler

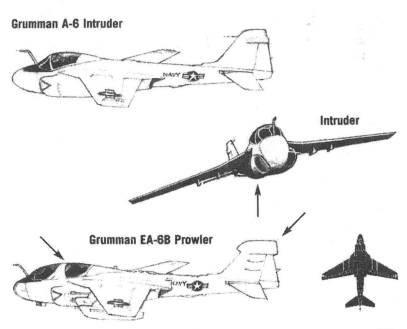

McDonnell Douglas AV-8B Harrier
Length: 46'4" (14.1 m) **Wingspan:** 30'4" (9.23 m)
Level flight: 655 mph (1,055 km/h) Mach 0.89 at sea level

Unmistakable; the Harrier has **huge air intakes extending from the wing root halfway to the nose**, giving it a very pointed nose when seen from below; a **bulbous canopy** makes it look a little more conventional from the side. **High, stubby wings carry four hardpoints**; these are noticeable even when the plane is not armed.

First flown at the Paris Air Show in the 1960s, a British Harrier I astonished the crowd by taking off vertically — the first fixed-wing attack aircraft in the world with no-runway, zero-roll capability. Now the workhorse of the U.S. Marines for forward air support, after a long struggle with the Navy, which wanted to keep combat support purely in Navy hands, and with Washington, where bureaucrats didn't want to import the Hawker Siddley (British Aerospace) design. A joint venture with McDonnell Douglas solved the problem.

Northrop F-5 Tiger II/Talon T-38 Trainer
Length: 46'–51' (14–15.5 m) **Wingspan:** 25'–26' (7.6–37.9 m)
Level flight: E version, 1,060 mph (1,706 km/h) Mach 1.6 at altitude

The T-38 version was used for 10 yr. by the U.S. Air Force Thunderbirds precision-flying team at airshows; the fighter-interceptor versions are very rare in the United States. The **small, oval engine intakes** and the simple, almost **triangular, wing and tail planes** are unique among military aircraft.

More than a thousand T-38s were used by the Air Force and Navy as trainers, and several thousand versions of the F-5 have been sold with Defense Department subsidies to non-Communist air forces throughout the world. A few remain in Navy and Marine service, where they play the role of aggressors in war games and gunnery exercises. Likeliest to be seen in southern California, going to and from military zones in the U.S. desert.

Lockheed-Martin F-16 Fighting Falcon
Length: 49'4" (15.03 m) **Wingspan:** 31' (9.45 m)
Cruising speed: Level flight: 1,300 mph (2,092 km/h) Mach 1.96 at altitude

Widely seen. The U.S. Air Force Thunderbirds have flown the F-16 since 1983. Head-on, note the "shark's mouth" air intake and the **drooping tail plane**; in side view, the plane appears to perch on top of the engine and shows a pair of **keel-like stabilizers** below the tail assembly; overhead, the **clipped triangular wing and tail planes** are diagnostic.

A bundle of graphite-epoxy wrapped around an afterburning turbofan jet engine, the F-16 started out as an experimental design to test lightweight construction techniques and ended up as the Air Force's choice as a combat fighting machine over battlefield areas. Since its adoption in 1975, the Air Force has turned it into a fighter-bomber and long-range interceptor, adding to its weight and cutting its maneuverability.

214

McDonnell Douglas AV-8B Harrier

Northrop F-5 Tiger

Talon T-38 Trainer

Lockheed-Martin F-16B

F-16

McDonnell Douglas–Northrop F-18 Hornet
Length: 65' (17.07 m) **Wingspan:** 37'6" (11.43 m)
Level flight: 1,190 mph (1,915 km/h) Mach 1.8 at altitude

One of the most easily identified of modern jet fighters: **half-round air intakes; twin tail fins lean outward; needle nose sweeps into a fairing into the wing; overhead, stubby, clipped, triangular wings and strongly swept tail planes.**

Because more news clips were broadcast from aircraft carriers than from land bases during the 1991 Gulf War, it sometimes seemed that the F-18 was the only U.S. fighter-bomber in the theater. Originally intended to be a single-seater, but two-seat trainers and fighter-bombers have been produced. Although designed for aircraft carriers, it is the fighter of choice in Canada (replacing F-101s, F-104s, and F-5s) and in Australia, where it is produced under license. Flown by the U.S. Navy's Blue Angels demonstration team.

Boeing (McDonnell Douglas) F-18E, F-18F Super Hornet
F-18E data: Length: 60' (18.3 m) **Wingspan:** 44'7" (13.6 m)
Speed: 1,221 mph (1,965 km/h) Mach 1.8 at altitude

The latest F-18 model is a new aircraft disguised to follow on in the F-18C and F-18D procurement program. The basic changes noted in the sketch are a much broader wing (the span is just a bit wider than the older F-18's), and the fairing to the wing has grown into an aerodynamic form rather than a simple streamlining. The new (for F-18s) square air intakes are derived from McDonnell Douglas's F-15 Eagle, but seen head-on, the planes can be easily differentiated: The F-18E and F have the splayed tail fins; the F-15's are vertical.

McDonnell Douglas–Northrop F-18 Hornet

Boeing F-18E Super Hornet

McDonnell Douglas F-15 Eagle
Length: 63'8" (19.42 m) **Wingspan:** 42'8" (13 m)
Level flight: 1,650 mph (2,655 km/h) Mach 2.5 at altitude

Increasingly common. **Massive rectangular engine air intakes,** wing and tail planes of **multifaceted geometry,** and **twin vertical tail fins.**

This airplane gives the impression of a great deal of mechanism crammed together. The small cockpit seems to bubble up higher and more abruptly than on any other modern jet fighter. A training version has two seats in tandem. The appearance of a large amount of engine and a small amount of airframe is indicative of the plane's performance: It is faster than all but the most advanced Russian MiG-25s at high speeds, and much more maneuverable than they are. May be seen with a bulge along the outside of each engine housing, indicating removable fuel tanks. These give the plane a maximum range of nearly 4,000 miles (6,437 km).

Grumman F-14 Tomcat
Length: 61'10" (18.85 m) **Wingspan:** fully spread, 64'1" (19.5 m); fully swept, 38'2" (11.63 m) **Level flight:** 1,560 mph (2,510 km/h) Mach 2.35 at altitude

A complex variable-wing plane. On first view, compare with the F-15 Eagle and the F-18 Hornet before deciding. On the flight line, **twin tail fins angle out slightly; rectangular air intakes angle inward at the top.** When the wings are extended at takeoff and landing, note the **bulky wing roots** housing the variable-geometry mechanism.

When the F-111 swept-wing proved much too heavy for carrier basing, the Navy chose the F-14 from a design competition. Separating Navy F-14s from Air Force F-15s by service markings will become increasingly difficult as planes are stripped of any distinctive painted markings that would make them identifiable on radar. F-15 Eagles have a smaller bubble canopy for a single pilot, whereas the F-14 carries a pilot and a radar-intercept officer under a longer canopy.

McDonnell Douglas F-15 Eagle

Grumman F-14 Tomcat

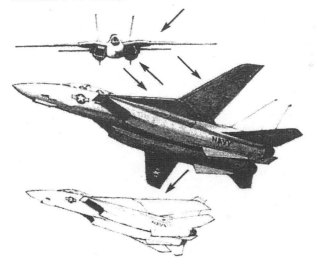

Lockheed Martin F/A-35
Length: 50'9" (15.48 m) **Wingspan** 43' (13.1 m) **Cruising speed:** 1,000 mph (1,600 km/h) estimated

An airplane characterized mainly by its **smooth exterior, a fuselage of curves** intended to scatter rather than reflect radar beams. If you took a model of a McDonnell Douglas F-18 Hornet and buttered it, it would resemble the F/A-35.

Once known as the Joint Strike Fighter, the F/A-35 is now intended to do everything from every conceivable launching platform. The STOVL (short takeoff, vertical landing) is to replace the VTOL (vertical takeoff and landing) Harrier AV-8 for the U.S. Marine Corps. The STOVL has a central fan that blows cool air straight down, with sufficient thrust to land the airplane softly, and with the help of the deflectable engine exhaust, let it take off from the Marine Corps' fleet of very short aircraft carriers or from a postage-stamp forward airfield. The Air Force will get a conventional-runway version with shorter wings and higher speeds, and the Navy will get a third variant, with either foldable wings or very stubby broad wings, that takes off and lands in the usual aircraft-carrier manner (slingshot takeoff assistance and tail-hook landings).

Lockheed Martin (with Boeing) F/A-22 Raptor
Length: 62'1" (18.92 m) **Wingspan:** 44'6" (13.56 m)
Cruising speed: in excess of 1,000 mph (1,609 km/h)

A plane of complex and irregular geometry: **Wings are slightly skewed truncated triangles** (approaching a delta wing style), the **paired tail fins are truncated right triangles,** and there is no word for the **irregular rhomboidal tail planes.** This aircraft, several years late and several billion dollars over budget, may be delivered to the U.S. Air Force in 2006.

The Raptor is an attempt at building an "air-dominance fighter" to replace the F-16, which was more modestly called an "air-superiority fighter." The Raptor, expected to begin operation in late 2005, is programmed to carry four medium-range air-to-air missiles (or two smart bombs) in the main weapons bay, as well as a pair of short-range air-to-air missiles in auxiliary bays on the sides of the fuselage. Each of its two engines has a parallelogram air intake at the wing root and an exhaust with an innovative two-dimensional thrust-vectoring system that can deflect the exhaust upward, pushing down on the aircraft's tail, accelerating the rate of change to a nose up (climbing) or deflect downward, pushing the tail up, sending the plane rapidly into a nose-down (diving) attitude. These exhausts can be independently operated to reduce or control roll.

Lockheed Martin F/A-35

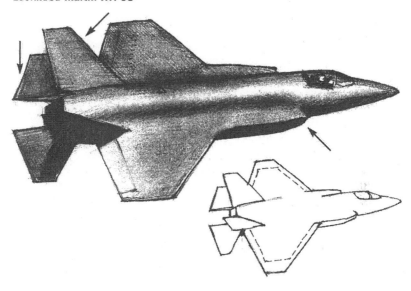

Boeing/Lockheed Martin F/A-22 Raptor

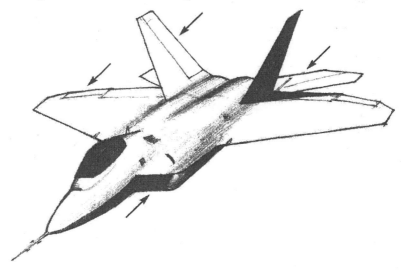

Lockheed S-3 Viking
Length: 53'4" (16.26 m) **Wingspan:** 68'8" (20.93 m)
Level flight: 506 mph (814 km/h) Mach 0.76 at altitude

Note the **twin jet engines pylon-mounted down and forward of the wing** and the **unswept wings;** overhead, it has noticeably **greater wingspan than length.**

When seen on alert, a long magnetic detecting boom extends to 15 ft. behind the tail. A carrier-based antisubmarine-warfare craft with a crew of four, it has the same mission as the land-based, turboprop Orion P-3 Electra. It is remarkably maneuverable for a reconnaissance aircraft, capable of dropping to sea level from 30,000 ft. in 2 min. In addition to magnetic detection, the S-3 has side- and forward-looking radar and infrared capacity. A few conversions to passenger and cargo uses for delivery to aircraft carriers are in service.

Lockheed U-2R, TR-1, ER-1
Length: 49'7" (15.11 m) **Wingspan:** 80' (24.38 m)
Cruising speed: 460 mph (740 km/h) Mach 0.69 at altitude

Very unusual configuration. **Single jet engine** and **80-ft. wingspan are unique. The sensor pods on the wings are integral,** not mounted on pylons. Some appear in civilian dress as environmental research (ER-1) aircraft. Mission pods vary.

The U-2, first flown in 1955, continues to be used as a platform for aerial observation from the ordinarily safe height of 80,000 ft. or more. In addition to the Air Force, NASA and other civilian agencies fly ER-1 for high-altitude scientific research. New versions, equipped with side-looking radar and laser equipment for selecting targets and guiding missiles and bombs to them, are designated TR-1. Large, wing-mounted fuel tanks give the U-2 the appearance of a twin jet when seen overhead. As scientific aircraft, they occasionally appear at civilian airports.

Lockheed S-3 Viking

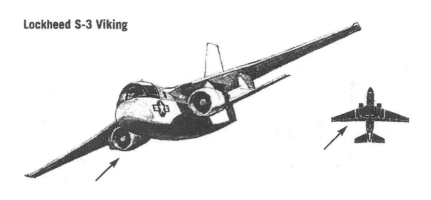

Lockheed U-2R

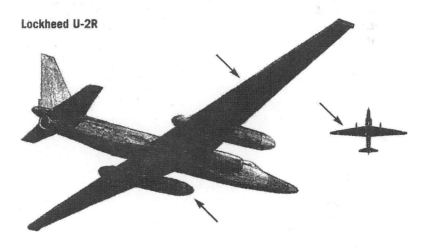

Lockheed C-5A Galaxy

Length: 247'10" (75.54 m) **Wingspan:** 222'8" (67.87 m)

Cruising speed: long range, 518 mph (833 km/h) Mach 0.78 at altitude

Uncommon. Compare with the C-141 StarLifter (next entry) before decid
ing. **Massive fuselage** with **high wing** and **T-tail. Four turbofan engines**
(noticeably larger in front, tapering to aft); overhead, compare with the
Boeing 747 silhouette (page 200).

The largest, and certainly the loudest, aircraft in North America, the C-5A
is an awesome sight on takeoff, with flaps fully extended and four engines
generating more than twice the noise of a Boeing 747. Viewed overhead, it
can be distinguished from the 747 (both have engines that taper noticeably
from front to back, unlike the C-141's) by the wing shape: There is very little
fairing, or widening, of the wing root on the C-5A as it enters the fuselage.

Lockheed C-141A StarLifter (and stretched C-141B)

Length: C-141A, 145' (44.2 m); C-141B, 168'4" (51.28 m) **Wingspan:** both models, 159'10" (48.74 m)

Cruising speed: 495 mph (796 km/h) Mach 0.75 at altitude

Based nationwide. On the ground, one of three **high-wing, four-jet,**
T-tail planes in North America. See the similar C5-A Galaxy and C-17 (previous
entry and next entry) for comparison. Confusing overhead, but the bulges
under and just aft of the **moderately swept wings** house the landing gear.

The Air Force's basic cargo and passenger aircraft, the jumbo-jet-sized
C-141 differs from all commercial four-engine jets by the combination of the
high wing and T-tail. Most C-141s have been stretched into the B versions,
which also have a domed fairing to house an in-flight refueling receptacle on
the top of the fuselage, just aft of the cockpit. Like many commercial jets,
the original C-141 had more lifting capacity than cabin capacity; the same
solution so common in airliners, stretching, though it improved total load
capacity, did not solve the problem created by the narrow cross-section of the
fuselage, which keeps it from carrying bulky items, such as full-sized tanks.

McDonnell Douglas C-17

Length: 174' (53.04 m) **Wingspan:** 169'10" (51.76 m)

Estimated speed: 520 mph (837 km/h) Mach 0.74 at altitude

Over a hundred C-17s are scattered from New Jersey to California; a total
of 180 will be delivered by 2008. **Huge, very fat fuselage; enormous, high**
tail fin is deeper at the top; upswept fuselage; large fairings under
wing for landing gear.

The C-17 is a cross, in many senses, between the short-field C-130
Hercules, which is too narrow for carrying outsized combat equipment, and
the wide-body C-5, which requires a long, paved runway. Design require-
ments are to carry, for example, three Bradley Fighting Vehicles or an M1
battle tank and support gear, with a maximum short-range payload of
172,000 lb. (78,108 kg), and to land that load on a forward airstrip only
3,000 ft. (914 m) long.

Lockheed C-5A Galaxy

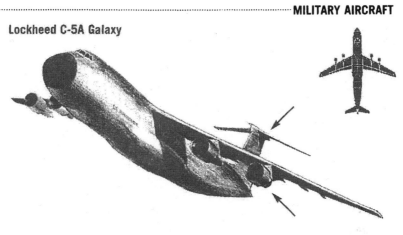

Lockheed C-141 StarLifter

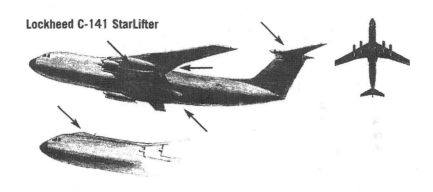

McDonnell Douglas C-17

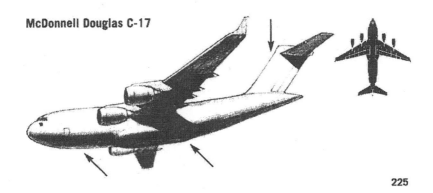

Boeing B-52 Stratofortress
Length: 157'7" (48 m) **Wingspan:** 185' (56.39 m)
Level flight: 650 mph (1,046 km/h) Mach 0.98 at altitude

Eight engines are carried in pairs below and forward of the wings' leading edges. Overhead, the contrails frequently show the eight exhausts, but note the unfaired swept wings and illusion of four engines; on the flight line, droopy winged.

Of the more than 550 B-52s built in the 1950s and early 1960s, a few hundred remain in service. Current models may show a bulge below the cockpit, housing forward-looking radar or low-light television. Many carry two air-to-surface missiles between the outboard engines and the wingtips. Many will be seen with a dozen wing-mounted, short-range Cruise missiles. Some current models may be carrying a number of wing-mounted rockets intended to divert heat-seeking surface-to-air antiaircraft missiles.

Rockwell B-1
Length: 143' (43.58 m) **Wingspan:** fully spread, 137' (41.75 m); fully swept, 78' (23.77 m)
Level flight: 1,454 mph (2,339 km/h) Mach 2.19 at altitude; subsonic at sea level

Huge, the size of a Boeing 707 or a stretched DC9 Super 80, with four engines mounted in pairs near the wing roots; wings extend for landing and takeoff, sweep back for operational flight; a sculptural quality to the drooping nose and fuselage-to-wing area; two beardlike winglets under the "chin" and a bulletlike "closeout" fairing to the tail end of the fuselage.

This plane is produced in small numbers but attracts attention by its size alone. You are unlikely to see it except with the wings fully extended unless you are near desert testing areas, where it will be executing supersonic, low-level maneuvers. On the ground, its massive, tall landing gear gives it a birdlike pose.

Boeing B-52 Stratofortress

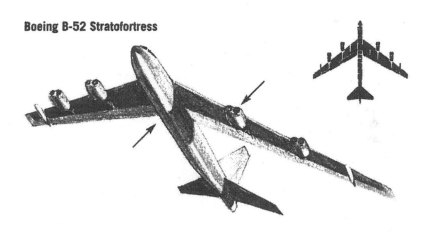

Rockwell B-1

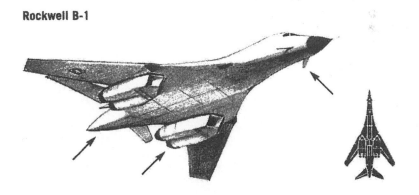

Lockheed F-117A Stealth Fighter

F-117 data: Length: 63'9" (19.43 m) **Wingspan:** 44'7" (13.6 m)
Speed: 685 mph (1,102 km/h) Mach 0.9 at altitude

A bizarre plane with **no curved surfaces anywhere.**

First, it isn't even a fighter. It's a very small, long-range, radar-evading bomber meant to carry a few smart bombs into heavily defended airspace. The plane is covered with a radar-absorbing material that is gooped on after construction and replaced frequently. The material gives the plane an odd optical effect: It will appear as any color, from reddish brown to neutral black. Used successfully for the first time in the 1991 Gulf War.

Northrop B-2 Spirit Stealth Bomber

Estimated data: Length: 69' (21.03 m) **Wingspan:** 172' (52.43 m)
Speed: 570 mph (1,010 km/h) Mach 0.84 at altitude

Some 20 of these flying wings are based in Missouri. They can also be seen routinely at their maintenance base in Oklahoma City and at Palmdale, California. Northrop made something of a specialty of building flying wings, starting with the pusher-propeller B-35 and the jet B-49 shortly after WWII. The B-2 is the only flying wing that became operational. The two air intakes each conceal a pair of jet engines, so the B-2 is actually a four-engined aircraft. It has enough range to fly from Missouri to Iraq and home again without refueling. There were no economies of scale; each B-2 costs at least $1.3 billion.

Lockheed F-117A Stealth Fighter

Northrop B-2 Spirit Stealth Bomber

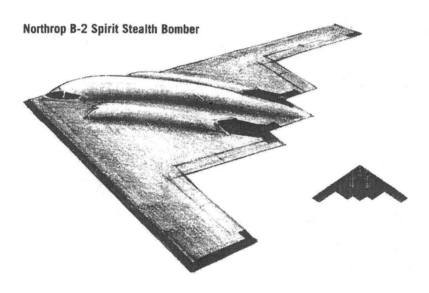

Rockwell OV-10 Bronco
Length: 41'7" (12.67 m) **Wingspan:** 40' (12.19 m) **Cruising speed:** 210 mph (338 km/h)

Overhead, the twin booms and **perfectly rectangular wing and tail plane** are diagnostic; near at hand note that the twin booms to the tail extend naturally out of the engine nacelles. The Cessna Skymaster is the only remotely similar aircraft.

The little OV-10 is a STOL (short takeoff and landing) observation and counterinsurgency aircraft that can operate without arresting gear from runways as short as the deck of a helicopter-carrying amphibious assault ship. A few heavily armed versions were in service with the U.S. Marines, including models for night observation: These have a distinctive probe, extending from the nose, that houses a forward-looking infrared sensor and laser used to guide missiles to the target. They were usually seen near bombing ranges, circling over practicing attack aircraft at a leisurely 55 mph (89 km/h).

Grumman OV-1 Mohawk
Length: 41' (12.5 m) **Wingspan:** 48' (14.63 m) **Level flight:** 289 mph (465 km/h)

Bulbous cockpit and triple tail give a sort of dragonfly look to the craft; wing tanks and a **right-side radar pod extend forward of nose.**

The Mohawk has such odd geometry that it can hardly be compared with any other aircraft. Although not all models have the curious radar pod that extends past the nose, the Grumman-style dihedral tail plane and triple tail fins are enough for positive identification. Most OV-1s carry two underwing fuel tanks just outboard of the engines. Only the Army flies the Mohawk, which is used as a target locater and battlefield mapper. The heavily armed Mohawks of the Vietnam War were refitted, as the Air Force, Navy, and Marines captured the fixed-wing attack plane mission from Army aviation.

Grumman S-2 Tracker, C-1 Trader, and E-1 Tracer
Length: 43'6" (13.26 m) **Wingspan:** 72'7" (22.13 m) **Cruising speed:** 150 mph (241 km/h)

Increasingly rare. In service as the Trader only, a shore-to-ship cargo plane; **twin engines that extend fore and aft of the symmetrically tapering wings; strong dihedral in tail planes.**

A typical Grumman aircraft. Note the bug-eyed cockpit (see the Mohawk, previous entry). When it was outfitted for advance warning of aircraft, it carried a teardrop-shaped radar dome 30 ft. long (compare with the current early-warning Hawkeye, with its round radar pod). Seen overhead, it could conceivably be confused with some commercial twin-engines, but the following combination is unique: symmetrically tapered wings; engine nacelles that extend well behind the wing; and a straight-line trailing edge on the tail plane.

Rockwell OV-1O Bronco

Grumman OV-1 Mohawk

Grumman C-1 Trader

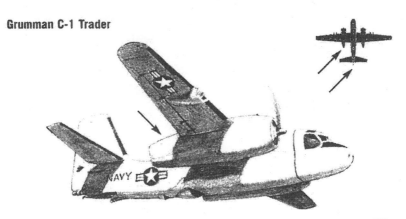

McDonnell Douglas A-4 Skyhawk
Length: 40' (12.2 m) **Wingspan:** 27'6" (8.38 m)
Level flight: 675 mph (1,086 km/h) Mach 0.89 at sea level

Almost extinct. Large engine air intakes sit above the wing roots; overhead, almost a delta wing look; refueling probe on starboard side of nose.

For years, the Navy's standard attack bomber, carrying more than 6 tn. armament (including nuclear bombs) on a relatively light 5 tn. airframe. The short triangular wings gave it carrier size without the complications of a folding wing and allowed for integral fuel tanks throughout the wings. Viewers of historical footage from the 1991 Gulf War will see modernized versions of the venerable Skyhawks (first flown in 1954) in the Kuwaiti air force.

Vought A-7 Corsair II
Length: 46'1" (14 m) **Wingspan:** 38'8" (11.78 m)
Level flight: 698 mph (1,123 km/h) Mach 0.9 at sea level

High-winged; large air intake and exhaust; overhead or on the ground, note the bulky fuselage without any apparent taper.

Once the Navy's standard attack bomber, based on the older and supersonic, now retired, U.S. Air Force F-8 design, bulked up for carrier duty. Like its World War II namesake, the old F4-U Corsair, it was a durable weapons platform with a long career. The Corsair II flew from Vietnam through the 1991 Gulf War; the original Corsair was in WWII and Korea and saw some duty in Vietnam.

McDonnell Douglas F-4 Phantom
Length: 58'–63' (17.7–19.2 m) **Wingspan:** 38'4" (11.7 m)
Level flight: up to 1,500 mph (2,414 km/h) Mach 2.25 at altitude

Drooping tail planes and upswept wingtips; overhead, a deep triangular wing and comparatively small tail plane.

A very large carrier-based aircraft, also once flown as a part of the U.S. Air Force. The fighter-bomber versions carried as much as 8 tn. munitions, more than the payload of a WWII B-29 Superfortress. It was once the basic interceptor, fighter-bomber, and electronic-reconnaissance aircraft for both the Navy and the Air Force, which accounts for the many nose configurations (see sketches). Still flown by several foreign air forces.

McDonnell Douglas A-4 Skyhawk

Vought A-7 Corsair II

McDonnell Douglas F-4 Phantom

nose variations

General Dynamics F-111, FB-111, and EF-111
Length: 73'6" (22.4 m) **Wingspan:** fully spread, 63' (19.2 m), fully swept, 31'11" (9.74 m)
Level flight: 1,650 mph (2,655 km/h) Mach 2.4 at altitude

On the ground, **thin swept wings** jut out of the **bulky wing roots** housing the variable-geometry mechanism; in side view, note a curious asymmetrical sculpting of the nose.

The F-111, developed as a supersonic fighter-bomber, evolved into a less common medium-range bomber (FB-111) and, in the EF configuration, as a radar suppressor and target locater. The rare EFs were distinguished by an electronic pod in the upper tail fin. What we have here is essentially a half-sized B-1 bomber (or perhaps the B-1 is an oversized F-111). Although one was unlikely to see an F-111 in the fully swept mode (the plane would be very high and going very fast), it would be distinguishable from delta wing planes by the notched effect where the wing meets the tail plane and by the clipped-off tail planes.

Douglas A-3 Skywarrior
Length: 76'4" (23.27 m) **Wingspan:** 72'6" (22.1 m)
Level flight: 610 mph (981 km/h) Mach 0.79 at sea level

Note the long, thin **swept wings with engines mounted well forward.** The wings enter the fuselage **without fairings.**

The A-3 was designed in 1952 as the first all-jet nuclear bomber to fly from a carrier deck and was the heaviest carrier-borne aircraft in any navy. But as bombs got lighter and aircraft more sophisticated, it was relegated entirely to mission support, either as a pure in-air refueling tanker or as a combination tanker–radar suppression plane.

General Dynamics F-111

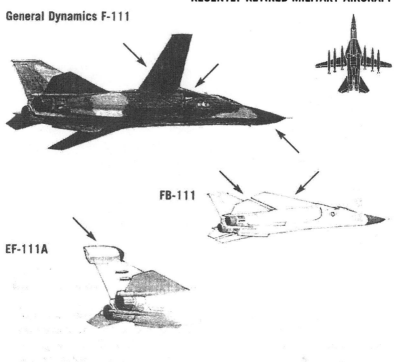

FB-111

EF-111A

Douglas A-3 Skywarrior

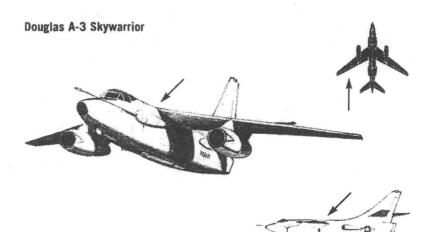

HELICOPTERS

Bell Model 47
Length: 43'7" (13.3 m) **Main rotor diameter:** 37'1" (11.32 m)
Cruising speed: 84 mph (135 km/h) **Useful load:** 1,025 lb. (416 kg) pilot and two passengers

For decades, this craft defined "helicopter." The huge one-piece bubble extends to the cabin deck, giving the craft a much more round-nosed look than the Lama/Alouettes it resembles.

When in military service, it was called the AH-1 Sioux; in its medical evacuation role, it starred in the opening sequence of the television series *M*A*S*H*. Various models were produced from 1945 to 1974, and many were converted for special uses, including the Continental El Tomcat agricultural sprayer, with a roll-bar cage cabin and a pointed nose replacing the Plexiglas bubble.

Aerospatiale SA-315B Lama
Length: 42'4" (12.91 m) **Main rotor diameter:** 36'1" (11.02 m)
Cruising speed: 75 mph (120 km/h) **Useful load:** 2,050 lb. (929 kg) crew of two and three passengers

One of two modern copters with the latticework of the tail boom exposed to view. The Lama (and earlier Alouettes) have a multipane bubbled canopy that stops well above the bottom of the cabin.

The most recent of a line of French-designed, Texas-built helicopters (it's basically an Alouette II frame with an Alouette III power plant), the Lama has a certified ceiling of 17,715 ft. (5,400 m) but has landed and taken off in the Himalayas at 24,600 ft. (7,500 m). It's a popular Alpine rescue helicopter and is named (using the French spelling) for the Andean two-*l* llama, not the one-*l* Tibetan priests.

Hiller UH-12
Length: 40'8" (12.41 m) **Main rotor diameter:** 35'5" (10.8 m)
Cruising speed: 90 mph (154 km/h) **Useful load:** 1,341 lb. (608 kg) crew of two and three passengers

Of all the slim-tailed small helicopters, only the UH-12 has a bent-up tail boom braced with a spar running from just below the main rotor to the top of the tail.

The original UH-12 (designated Hiller 360) was the military's H-23 Raven. Out of production for a decade, it has been revived, and many UH-12s now have turboshaft engines. The UH-12 is used primarily as an agricultural sprayer and seeder.

Bell Model 47

Aerospatiale SA-315B Lama

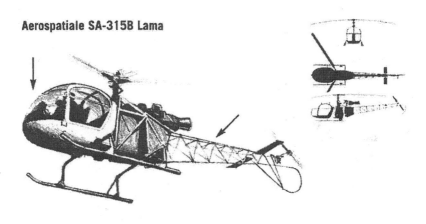

Hiller UH-12

Schweizer Model 300 (Hughes 269/300)

Length: 30'10" (9.4 m) **Main rotor diameter:** 26'10" (8.18 m)
Cruising speed: 77 mph (124 km/h) pilot and two passengers

Oval windows in the doors give this aircraft the most **dragonfly-looking** head of all helicopters, especially when combined with the **slim tail boom braced from beneath the fuselage.**

Developed in the 1950s, the Hughes 269/300 was used as the TH 55A Osage, the U.S. Army's basic helicopter trainer. The Schweizer Corporation has concentrated, with much success, on supplying police versions with some armor plating and special floodlighting and public-address systems. The slightly larger Schweizer 330 shows a small window behind the door, making it look like what it is: something halfway between a Schweizer 300 and the Hughes/McDonnell Douglas 500.

Robinson 22, R22 Beta II

Beta II data: Length: 21'7" (6.6 m) **Main rotor diameter:** 25'2" (7.68 m)
Cruising speed: 110 mph (177 km/h) **Useful load:** 538 lb. (244 kg)

Robinson helicopters all have a **proportionally very tall main rotor mast,** combined with a **very long, very thin boom** to the tail rotor. The engine is exposed just rear of the cabin.

The R22 was the "pony car" of the helicopter world: small, inexpensive, fun to fly, uncomplicated to maintain (for a helicopter). Very popular as trainers and in roles that really required only a pilot. Seen at seaports with floats, they are used to spot tuna and swordfish. The larger R44, next entry, brought Robinson into the larger market of law enforcement and related activities.

Robinson R44 Raven

Length: 29'5" (8.96 m) **Main rotor diameter:** 33' (10.06 m)
Cruising speed: 120 mph (194 km/h) **Useful load:** 774 lb. (350 kg)

From any side angle, Robinson helicopters show an **extremely tall main rotor pylon;** a truly **long, tapering tail boom;** and a **simple low-dipping two-piece windshield** but without the typical down-to-the-floor window for the pilot. The four-seat 44 shows two side windows, each side, which are also doors, for a total of a door for every person aboard. Unlike the Robinson 22 (previous entry), the **44's engine is enclosed.**

One of the fastest-selling helicopters in history, with specific models factory-built as police, news camera, search and rescue, and float-craft. The base price has been kept below $300,000 for the R44 since its inception in 1999. With an old-fashioned Lycoming piston engine that can be worked on at any airfield in the world, it is the flying equivalent of a nice, sturdy, car-bureted truck that you can tune with a screwdriver and a strobelight.

HELICOPTERS

Schweizer Model 300 (Hughes 269/300)

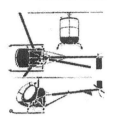

Robinson 22

Robinson R44 Raven

239

Schweizer Model 330, 330SP, 333

333 data: Length: 22'5" (6.82 m) **Main rotor diameter:** 27'6" (8.39 m)
Cruising speed: 120 mph (192 km/h) **Useful load:** 1,340 lb. (590 kg) pilot and four passengers

Compare with the MD 500 (next entry) before deciding: **Fuselage sweeps up to the tail assembly; a pair of horizontal stabilizers on the fuselage** (not on the top of the vertical stabilizer, as in the MD 500). **The windscreen is even more buglike,** with the only side window reduced to a smidgen.

In the 1990s, Schweizer created the fuselaged (not boomed and braced like their original 300) Model 330; the latest 33 is an upgrade involving both new and more efficient rotor blades and a larger rotor disk to handle the increased torque. An upgrade kit with these features has been used on both late-model 300s and 330s, further confusing the issue. Ancestry is oblique: Hughes designed the first 300s and 500s; McDonnell Douglas turned that Hughes program into the MD 500 (next entry). Schweizer then adopted the McDonnell Douglas MD 500 cockpit design for its 330s. All members of the Hughes/McDonnell Douglas/Schweizer programs are widely in service in police and other surveillance roles.

McDonnell Douglas MD 500, 530 (Hughes 500)

MD 500 data: Length: 30'10" (9.4 m) **Main rotor diameter:** 26'4" (8.03 m)
Cruising speed: 137 mph (220 km/h)
Useful load: 1,559 lb. (707 kg) various passenger loads; maximum, pilot and six passengers

Distinctive **sharp-nosed multipane canopy, glass to deck; irregular large windows in each of four doors.** Earlier Hughes models had a rounded nose but similar door-window treatment. Later 500 and 530 models have a **T-tail horizontal stabilizer above the tail rotor.** The long, slim boom is fully faired into the fuselage.

Widely seen as a commuter carrier and executive aircraft in the United States. Versions of the basic MD 500 are in military service in several foreign countries, marketed with TOW (Tube-launched, Optically tracked, Wire-guided) missile mounts as the Defender. The earlier Hughes 500 was the U.S. Army's OH-6 Cayuse.

Schweizer Model 333

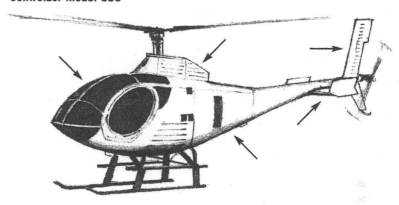

McDonnell Douglas MD 500, 530 (Hughes 500)

McDonnell Douglas Helicopters MD 520N

Length: 25'6" (7.77 m) **Main (and only) rotor diameter:** 27'4" (8.33 m)
Cruising speed: 150 mph (240 km/h) **Useful load:** 1,765 lb. (800 kg) crew of two

One of three MD Helicopters (see also next two entries) characterized by a nontapering cylindrical tail boom and with no tail rotor. The MD 520N is basically the older MD 530 cockpit (page 240) with the new NOTAR tail system. Close at hand, note a grilled air intake (looks like a rear-view window) behind the cockpit and, if it is turned toward the viewer, a slotted vent in the extreme end of the fuselage, just astern of the twin-tail.

The NOTAR system, devised by Boeing, is used in MD Helicopters' latest models. Briefly, an internal fan, driven by the main engine, pushes air down the symmetrical tail boom. The fan first vents down from two slots in the bottom of the boom, causing the downdraft from the main rotor to flow evenly across and down the port side of the boom, which gives a lateral lift, as though it were moving over an airplane wing. This lateral lift provides the first resistance to rotation of the fuselage about the main rotor drive shaft. In addition, a single vent at the end of the tail boom can be rotated to direct a thrust of air in the needed direction. The combination of the two air effects substitutes a very simple mechanical system for the complexities (and disadvantages) of a tail rotor.

McDonnell Douglas Helicopters MD 600N

Length: 30'6" (9.3 m) **Rotor diameter:** 27'6" (8.38 m) **Cruising speed:** 150 mph (240 km/h) **Useful load:** 2,000 lb. (907 kg) crew of one and up to seven passengers.

Essentially a stretch of the MD 520N (previous entry). **Three side windows, NOTAR air intake just behind the engine, twin tail (slight variations are out there), and no tail rotor. Air jet at extreme end of tail boom** may be visible; see previous entry for explanation of NOTAR.

If you see a helicopter patrolling the Mexican border, it's likely to be an MD 600N; the twin tail (resembles the old airplane style, see page 135) will be definitive. Both larger and more powerful than its parent MD 520N, the MD 600N features an unusual six-bladed rotor, noticeable if you find one parked nearby. The extra lift has made it a popular choice in hot climates: not only the Mexican border but also for sightseeing in the Grand Canyon.

MD 520N

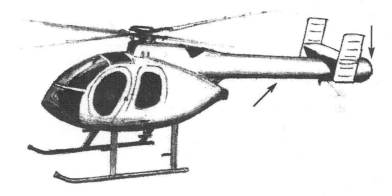

MD 600N

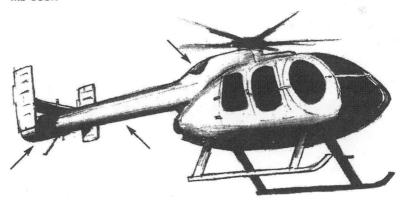

MD Helicopters MD Explorer (MD 900, 902)

Length: 32'8" (9.86 m) **Rotor diameter:** 33'10" (10.34 m) **Cruising speed:** 155 mph (250 km/h)
Useful load: 1,601 lb. (726 kg) pilot and up to seven passengers

By far the bulkiest of the MD tail-rotorless helicopters, two fully enclosed turboshaft engines that appear piled on top of the fuselage show an unusual side air intake. Air intake between engines for the NOTAR system is almost always obscured; rotatable air jet at the end of the tail cone may be visible.

When it acquired McDonnell Douglas, Boeing had to spin off the McDonnell Douglas light-helicopter division; a sale to Bell Textron was vetoed by the U.S. government for fears of monopoly. MD Helicopters is indirectly owned by a European consortium, RDM, which believes that the modest improvements made over the original McDonnell Douglas MD 900 are of European innovation.

Brantly-Hynes B-2, Brantly B-2B

Length: 28' (8.53 m) **Main rotor diameter:** 23'99" (7.24 m) **Cruising speed:** 90 mph (145 km/h)
Useful load: 610 lb. (276 kg) dual-control two-seater

The Brantly trademark is the long tail that grows out of the cockpit like an aluminum ice-cream cone. The odd little half-round windows in the doors are also unique.

This is the little helicopter that could; certified in 1959, it has been produced by half a dozen companies, and the type certificate is now owned by a Chinese company that has a new manufacturing plant at Vernon, Texas. Priced recently at $150,000, it is the low end of the rotorcraft market. Sometimes seen on floats and not an uncommon agricultural sprayer.

Brantly-Hynes 305

Length: 32'11" (10.03 m) **Main rotor diameter:** 28'8" (8.74 m)
Cruising speed: 110 mph (177 km/h) **Useful load:** 900 lb. (408 kg) crew of two and three passengers

This is the jumbo ice-cream cone, with odd-geometry passenger windows showing behind half-round door windows. Standard model delivered with unusual fixed tricycle gear, but even on floats, it would look like a Brantly-Hynes.

A few dozen B-H 305s survive in the United States and Canada, most in corporate executive service. International travelers may remember them as airline connectors at Heathrow, England.

MD Explorer MD 900

Brantly-Hynes B-2

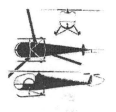

Brantly-Hynes 305

ENSTROM F28F, FALCON F28F

Most recent models of the evolving Enstrom two-passenger helicopters show minor but noticeable differences from the earlier aircraft. The F28 Falcon (upper sketch) has added end plates to the horizontal stabilizer, upgraded to a one-piece windscreen, and, distinctively, placed the drive shaft to the tail rotor outside the tail cone. The F28F (lower sketch) kept the open cable tail rotor guard, the end-plated horizontal stabilizers, and the one-piece windscreen but put the drive shaft back into the tail cone.

Enstrom 480B

Length: 28'10" (9.09 m) **Main rotor diameter:** 32' (9.75 m)
Cruising speed: 126 mph (202 km/h) **Useful load:** 1,305 lb. (592 kg) pilot and three passengers

The 480 is very similar to the Enstrom 28F series (preceding entry) but has added **end plates on the horizontal stabilizer,** substituted **simple wide struts to the skids** (while dropping the little rear wheels), and also added a **single, narrow vertical window behind the door** for the passengers (the "2" in the 280 and 28 meant two people sitting side by side; the "4" in 480 signifies the cabin enlargement to four places). Rather **long shaft to the main rotor.** Flying by, you may notice the different sound created by the 480's turboshaft engine, compared to the original 280's piston engine.

Enstrom is in some sense the Model A Ford manufacturer of helicopters, providing a serviceable helicopter at a reasonable price. The first 480s sold for much less than a million dollars.

Enstrom 280FX, 280FX Shark

280FX data: Length: 29'4" (8.7 m) **Main rotor diameter:** 32' (9.75 m)
Cruising speed: 107 mph (172 km/h) **Useful load:** 1,015 lb. (460 kg) pilot and one passenger

Most recent models of the evolving Enstrom two passenger helicopters show minor but noticeable differences from the earlier aircraft. The F28 Falcon has added end plates to the horizontal stabilizer, upgraded to a one-piece windscreen and, distinctively, placed the drive shaft to the tail rotor outside the tail cone. The 280FX has adopted the open cable tail rotor guard, and added the end-plated horizontal stabilizers and the one-piece windscreen. The 280 FX Shark, if you see one parked next to plain old 280 FX, has a longer cabin area and some subtle streamlining. Like all the Enstroms, they have a kind of unpretentious sleekness.

Enstrom, a small company founded by the aviator and inventor Rudolf Enstrom, has produced over 1,000 helicopters and has been owned at different times by two corporations and a group of private investors, including F. Lee Bailey, the celebrity lawyer.

Enstrom F28 Falcon

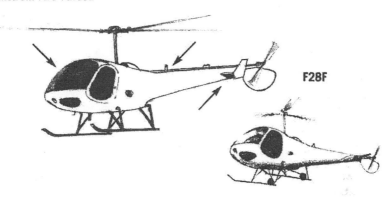

F28F

Enstrom 480B

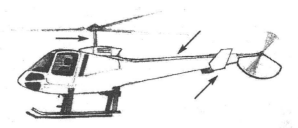

Enstrom 280FX Shark

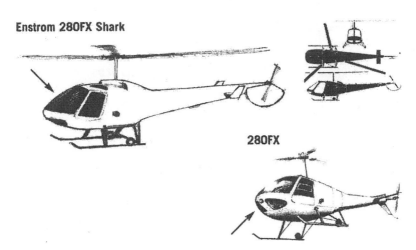

280FX

Hiller FH-1100 Pegasus
Length: 41'3" (12.57 m) **Main rotor diameter:** 35'5" (10.8 m) **Cruising speed:** 122 mph (196 km/h)
Useful load: 1,030 lb. (468 kg) pilot and four passengers

Not common. Low, almost horizontal tail boom emerges from the bottom of the bulky fuselage. A large L-shaped exhaust shows below the main rotor, and the engine location is clearly visible, above the fuselage.

This infrequently encountered craft was developed for a military competition — a fly-off for an observation and light-transport vehicle. The winner was the Vietnam-era Hughes OH-6 Cayuse (unofficially named the Loach in-country; see McDonnell Douglas/Hughes 500 [page 240]). Although much modified since the original military model, the unique overall shape of the Pegasus is unchanged.

Aerospatiale Alouette III
Length: 42'1" (12.84 m) **Main rotor diameter:** 36'1" (11.02 m) **Cruising speed:** 122 mph (197 km/h)
Useful load: 2,386 lb. (1,078 kg) pilot and up to seven passengers

Uncommon. Note the very large multipane greenhouse canopy. Exposed turboshaft engine. Upside-down tail rotor guard also serves as a landing skid. Standard with nonretractable tricycle landing gear.

Astonishing high-altitude performance makes this a popular police and rescue craft in the European Alps and a small workhorse load lifter in the American Rockies. With a crew of two and a 550 lb. (250 kg) payload, this copter took off and landed at 19,698 ft. (6,004 m) in the Himalayas.

MBB/Kawasaki BK 117
Length: 42'8" (13 m) **Main rotor diameter:** 36'1" (11 m) **Cruising speed:** 158 mph (264 km/h)
Useful load: 1,948 lb. (917 kg) pilot and seven passengers

Not common but noticeable and distinctive. Large pod fuselage, high horizontal tail boom, huge angled vertical stabilizers. Also note the large air intake and visible exhaust pipe just below the always horizontal rigid rotors.

This is a joint German-Japanese venture: MBB providing the running gear; Kawasaki, the airframe and electronics. Similar to the MBB 105 but with larger engines, more capacity, and higher speed. A few of the first 100 produced are in air ambulance service in the United States and attract attention by their unusual speed.

Hiller FH-1100 Pegasus

Aerospatiale Alouette III

MBB/Kawasaki BK 117

MBB BO 105
Length: 38'11" (11.86 m) **Main rotor diameter:** 32'3" (9.83 m) **Cruising speed:** 127 mph (204 km/h)
Useful load: 2,425 lb. (1,100 kg) pilot and four passengers

Combines a deep fuselage with a short-looking horizontal tail boom set high under the main rotor; large air scoop in front of the rotor drive shaft. Rear passenger-door window is much smaller than the one in the pilot's door. The largest model (LSB, illustrated) has a cabin that is a foot longer than standard and has a third small side window. Small end-plate fins on horizontal stabilizer. When the aircraft is stopped, note the non-drooping, rigid rotors.

This German invention (MBB stands for Messerschmitt-Bolkow-Blohm) is manufactured in Canada and in Germany. The rigid main rotors can be pitched to push the craft down (other helicopters can rise or sink but not achieve negative gravity), making it capable of ground-hugging flight in combat, less sensitive to downdrafts, and quicker when moving from assigned altitude to ground level. Many in air ambulance and search-and-rescue use.

Eurocopter AS350 (Aerospatiale Ecureuil 350, briefly Astar in North America)
Length: 42'7" (12.99 m) **Main rotor diameter:** 35' (10.68 m) **Cruising speed:** 144 mph (232 km/h)
Useful load: 1,847 lb. (838 kg) pilot and five or six passengers

One of the commonest large utility civilian helicopters, with more than 350 in the United States. Pointy nosed; teardrop look-down windows like nostril openings; enclosed engine; up close, the tail rotor drive-shaft housing lies on top of the tail cone. The Twinstar is similar; shows two air intakes and exhausts for its paired engines.

Current production is Ecureuil and Twinstar worldwide, as the Lycoming-powered Astar has been superseded by craft with French-built Turbomeca engines. Aerospatiale builds flyable helicopters in France, where they are test-flown in a "green" state, then disassembled, shipped, and reassembled in Texas, where final wiring, piping, and all avionics are installed, along with customizing details.

Aerospatiale Gazelle
Length: 39'3" (11.97 m) **Main rotor diameter:** 34'5" (10.5 m) **Cruising speed:** 144 mph (233 km/h)
Useful load: 1,460 lb. (661 kg) pilot and four passengers

Classic European design with the tail rotor enclosed in the tail fin (in a Fenestron). The Gazelle has a huge bubble-fronted greenhouse canopy; enclosed turboshaft engine stands out behind main rotor drive shaft. Near at hand, tail rotor drive shaft lies on tail cone. Has a rigid, non-drooping main rotor similar to the MBB series (page 250).

Quite rare in the United States; many more in French and British military service. No longer produced in North America.

MBB BO 105

Aerospatiale Ecureuil 350

Aerospatiale Gazelle

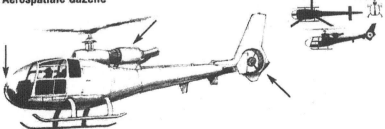

Eurocopter EC 120 Colibri Hummingbird

Length: 31'6" (9.6 m) **Main rotor diameter:** 30'6" (9.3 m) **Cruising speed:** 119 mph (191 km/h)
Useful load: 1,664 lb. (755 kg) pilot and three passengers

The cockpit glass is almost a wraparound, with single knee-high upcurve of fuselage in front of the pilot's seat, something like a bib; the EC 130, next entry, has a pair. The drive shaft to the Fenestron is enclosed within the tail boom, and the skid, similar to the one on the EC 130, has a unique wide rear strut on each skid.

This is a downscale stablemate of the EC 130 and has made a few sales to U.S. police departments. It is being assembled in China, as well as in the United States and Europe.

Eurocopter EC 130

Length: 35' (10.68 m) **Main rotor diameter:** 35' (10.68 m) **Cruising speed:** 146 mph (235 km/h)
Useful load: 2,293 lb. (1,040 kg) pilot and six passengers

One of a pair of similar new Eurocopters: the EC 130 and its skinnier brother, the EC 120 (previous entry). The cockpit window is unique to both: **Windscreen is a floor-to-ceiling bubble of Plexiglas in each; the EC 130's two extensions of the fuselage, directly in front of the pilot and adjacent passenger, almost look as though the cockpit has a brassiere on. The drive shaft to the Eurocopter-trademark enclosed tail rotor (Fenestron) runs through a housing outside and on top of the tail boom.**

Relatively new to the market and not as widespread as other Eurocopters, the EC 130 offers excellent vision for all aboard, critical to police and other agencies with surveillance responsibilities.

Eurocopter EC 135

Length: 33'6" (10.2 m) **Main rotor diameter:** 33'6" (10.2 m) **Cruising speed:** 150 mph (241 km/h)
Useful load: 2,789 lb. (1,265 kg) pilot and five or six passengers

Quickly identifiable as a Eurocopter by the **Fenestron (enclosed) tail** rotor and likelier to be seen with corporate logos or hospital markings than with police identity. The **fuselage is distinctly egg shaped; a pair of up-and-down swept end plates on the horizontal stabilizer. Always with skids.**

The EC 135 has been very successful in Europe as a police and border patrol helicopter, but of the first 250 built, only a handful are in U.S. police fleets. The planes are delivered with a variety of engine options; model numbers include the first initials of Pratt & Whitney, and Turbomeca (an R for Rolls-Royce may appear sometime in the future).

Eurocopter EC 120

Eurocopter EC 130

Eurocopter EC 135

Aerospatiale Dauphin II, USCG HH-65 Dolphin

Length: 45'6" (13.88 m) **Main rotor diameter:** 39'2" (11.94 m) **Cruising speed:** 160 mph (257 km/h)
Useful load: 4,341 lb. (1,969 kg) crew of 2 and 11 passengers

In its Coast Guard red and white, one of the most common coastal helicopters. The **tall, large tail fin encloses Fenestron tail rotor, fully retractable gear, and end-plated horizontal stabilizer forward of tail fan rotor.** Civilian versions show various window designs, but the tail end is diagnostic.

Like other Aerospatiale craft, about 40% of the value is added in the Texas assembly plant, but it still took a wonderful design to make this the first foreign helicopter to win a U.S. government competition. In addition to Coast Guard short-range short-based rescue, Dauphin/Dolphins are carried on icebreakers and cutters. Executive versions take advantage of the large cabin to build a sound cocoon that depresses ambient noise to a level equivalent to highway noise inside a luxury sedan.

Agusta A109 Hirundo

Length: 42'9" (13.05 m) **Main rotor diameter:** 36'1" (11 m) **Cruising speed:** 144 mph (233 km/h)
Useful load: 2,600 lb. (1,180 kg) pilot and seven passengers

One of the few truly streamlined helicopters. **Fully retractable landing gear; slim nose; swept-back tail fins; engine, smoothly wrapped, sits just under main rotor.** Compare with Sikorsky Spirit (next entry).

One of Italy's most successful aircraft, the Hirundo carries the president of Italy and dozens of U.S. corporate executives. *Hirundo* is "swallow" in Italian, and this helicopter moves with comparable rapidity. Originally single engined, everything you see will be twin engined, for emergency reliability.

Sikorsky S-76 Spirit

Length: 52'6" (16 m) **Main rotor diameter:** 44' (13.41 m) **Cruising speed:** 144 mph (232 km/h)
Useful load: 4,700 lb. (2,132 kg) crew of 2 and up to 12 passengers

Huge but still streamlined. The Spirit's **tall, slim tail fin is all above tail cone; twin engine air intakes obvious below main rotor; odd-geometry side windows; very unusual paired look-down windows;** fully retractable landing gear.

Sikorsky designed this craft specifically for the support of offshore oil platforms; it is capable of ferrying rig workers and equipment for considerable distances. Many have been modified for air ambulance work, with redundant backup electrical service, suction, and medical gases delivery. Extreme utilization of lightweight materials is typical, and many modifications for increased quiet and dampened vibration have been incorporated since 1980.

Aerospatiale Dauphin II, USCG HH-65 Dolphin

Agusta A109 Hirundo

Sikorsky S-76 Spirit

AgustaWestland A119 Koala
Length: 36'4" (11.07 m) **Rotor diameter:** 35'6" (10.83 m) **Cruising speed:** 155 mph (250 km/h)
Useful load: 2,910 lb. (1,320 kg)

This derivative of the Agusta A109 (page 254) shares a common characteristic: **The tail rotor is centered on the vertical tail assembly and revolves within the limits of the dorsal and ventral fins. The housing for the engine and main rotor base is relatively streamlined.** Although it has just one huge turboshaft engine, it sports the dual exhausts from the twin-engined A109. The centered tail rotor is very European; see the otherwise dissimilar Aerospatiale Ecureuil 350 (page 250).

AgustaWestland is a merger of the Italian Agusta with the British Westland, and this plane was their first teamwork project. Koalas remain rather rare, having unfortunately come on the market just as the Internet bubble burst in 2000. Westland's military helicopters were brutes, but the Agusta-Westland products carry not a hint of the British company's design style.

Bell 206 JetRanger, LongRanger (OH-58 Kiowa), Bell 407
JetRanger data: Length: 38'9" (11.82 m) **Main rotor diameter:** 33'4" (10.16 m)
Cruising speed: 133 mph (214 km/h) **Useful load:** 1,745 lb. (791 kg) pilot and four passengers

Very Bell looking, with its **dorsal and ventral tail fins and tail plane midway down high-set tail boom.** The seven-passenger LongRanger adds a third side window and end plates on the tail plane/horizontal stabilizers (see sketch).

Designed as a military aircraft, but thousands of the civilian versions are in service as ambulances and in police, traffic reporting, and news-gathering roles. Ordered by the military in 1968 after considerable delays and price overruns by the original competition winner, the Hughes OH-6. The LongRanger has been updated and restyled Bell Helicopter 407.

Bell 222, 230, 430
Length: 50'4" (15.36 m) **Main rotor diameter:** 42' (12.8 m) **Cruising speed:** 161 mph (269 km/h)
Useful load: 3,350 lb. (1,520 kg) pilot and seven passengers

An aircraft with a number of bumps and projections: **Long, pointy nose; retractable landing gear in pods (sponsons) that jut straight out; engine exhausts project out to the side; tail plane sticks out halfway down the tail cone; dorsal and ventral tail fins; odd-geometry side windows.**

The first twin-engined civilian helicopter built in North America at Bell Textron's Quebec plant. The large sponsons holding the main landing gear act as an additional airfoil and provide some lift at higher forward speeds. The main changes have been in engines; the latest, 430, has Rolls-Royce turboshafts driving the typical modern four-bladed propeller.

AgustaWestland A119 Koala

Bell 206 JetRanger

LongRanger

Bell 222

Bell 427
Length: 36' (10.98 m) **Main rotor diameter:** 36'10" (11.22 m)
Cruising speed: 155 mph (250 km/h) crew of two and six passengers

Take the Bell 206 Kiowa (page 256) and make it a **twin turboshaft with the engines fully enclosed** and with **prominent angled-out exhaust tubes each side.**

Bell took the successful Kiowa airframe with the LongRanger end-plated horizontal stabilizers (page 256) and added the second engine for more reliability and also got a few more miles per hour in the bargain.

Bell 212, 412, "Huey"
412 data: Length: 56' (17.07 m) **Main rotor diameter:** 46' (14.02 m)
Cruising speed: 161 mph (269 km/h) **Useful load:** 4,233 lb. (1,920 kg) crew of 2 and up to 12 passengers

Most modern of the Bell 204/UH-1 Huey series; see additional side-view silhouettes for details. The 412 shows a **four-bladed main rotor and well-streamlined engine housings.**

The 212/412 series started with the old model 204 UH-1 Huey, which grew into the longer, slimmer model 205 UH-1 Iroquois. The big change to 212 added twin engines for redundant reliability and increased performance. The 412's four-bladed rotor increased performance and decreased fuselage vibration. Versions are manufactured in several countries. Armed Hueys and Iroquois are rare now; many have been upgraded and converted for military medical evacuation and light transport. Latest model, the Huey II, is most easily distinguished by counting the windows: It has **two side windows;** the smaller window behind the cockpit has been eliminated.

Bell 214 ST
Length: 62'2" (18.95 m) **Main rotor diameter:** 52' (15.85 m) **Cruising speed:** 161 mph (295 km/h)
Useful load: 4,233 lb. (1,920 kg) crew of 2 and 18 passengers

Huge and pointy nosed; four side windows; tail rotor mounted on top of tail; long engine fairing with large air scoops and exhausts.

The 214 was designed for the Shah of Iran as a troop transport to be built under license in Iran. The fundamentalist revolution intervened, and the 214's actual first use was as a petroleum-platform service bus in the North Sea. May be delivered with wheels instead of skids (for airport transportation) and has been provided with optional deicing apparatus for Arctic work, including the Alaskan North Slope oil field.

Bell 427

Bell 212, 412 Huey

Bell 205, UH-1 Iroquois

Bell 204, UH-1 Huey

Bell 214 ST

Bell Agusta AB139

Length: 51' (15.53 m) **Main rotor diameter:** 51'10" (13.8 m) **Cruising speed:** 180 mph (290 km/h)
Useful load: 5,500 lb. (2,495 kg) crew of 1 or 2 and up to 15 passengers

Still quite rare. Overall, a pleasing symmetry: horizontal control surfaces each side, exhaust for twin turboshaft engines melded into the fuselage, elegant look-down cockpit windows, large sliding doors on both sides. Part of the retractable gear element creates a passenger step, each side.

Bell came quite late into this project; the basic design is entirely from AgustaWestland. This aircraft does pose competition to Bell's very successful (and rather elegantly designed) 222 (page 256). With greater lift and endurance, the AB139 has attracted buyers in New Zealand, where helicopters are a way of life in the Southern Alps.

Eurocopter (formerly Aerospatiale) 322 Super Puma

Length: 61'4" (18.7 m) **Main rotor diameter:** 51'2" (15.6 m) **Cruising speed:** 165 mph (266 km/h)
Useful load: 9,128 lb. (4,140 kg) crew of 2 and up to 21 passengers

Uncommon in North America. Note the twin engines mounted well forward, almost in line with main rotor shaft; vented, slatted horizontal stabilizer on port side; large ventral fin/tail skid. Also, retractable tricycle landing gear and an additional, much larger side window than on AS 330 Puma.

More common in Western Europe, where it is in military, VIP, and commuter airline use. The Super Puma was the first helicopter certified to fly into known or predicted icing conditions. For that reason, it is a common sight at ports surrounding the North Sea petroleum fields, serving as a platform-crew airbus.

AgustaWestland EH101, Canadian Defense Force CH-149, US101

Length: 64'5" (19.63 m) **Main rotor diameter:** 61' (18.6 m) **Cruising speed:** 168 mph (270 km/h)
Useful load: 13,228 lb. (6,000 kg) pilot and up to 30 passengers in sardine mode

Rare in North America but soon to be highly visible. Twenty-three were ordered for the presidential fleet (as US101) in 2005 and should be taking off for Camp David in 2009. This helicopter shows three prominent engine exhausts (unique among all copters), five congruent passenger windows on each side, and a bubble in the door in the search-and-rescue models. It slightly resembles another very large European machine, the Eurocopter (Aerospatiale) 322 Super Puma (previous entry).

The EH101 is only the second helicopter, after the Super Puma, to be certified for flying into known icing conditions, one of the reasons for its adoption by the Canadian, British, and Danish defense forces. It has not yet cut into the North Sea oil platform role captured by the Super Puma.

Bell Agusta AB139

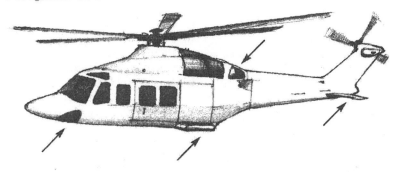

Eurocopter (Aerospatiale) 322 Super Puma

AgustaWestland EH101

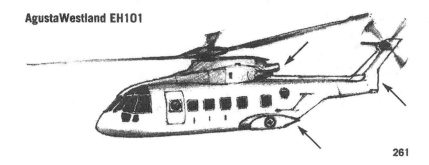

Bell OH-58D Kiowa, SeaRanger
Length: 40'11" (12.49 m) **Main rotor diameter:** 35'4" (10.77 m)
Cruising speed: 117 mph (188 km/h) **Useful load:** 1,736 lb. (696 kg) crew of two

It looks like a Bell 206 JetRanger that had a bad experience in a body shop: Flat platform tops a massive engine housing; mast-mounted "eyeball" gun sight typical; armament sponsons project below cockpit.

Technically, it's a scout helicopter, but it can be armed with air-to-air or air-to-ground missiles and multiple machine guns, all sighted through the mast-mounted television eye, built for normal optical or infrared imaging. Prior to the invasion of Kuwait, used in the Gulf War to suppress gunboat attacks on oil tankers. Rarer than the Apache or HueyCobra attack helicopters.

Bell AH-1 HueyCobra, SeaCobra, Super Cobra
HueyCobra data: Length: 53'1" (16.18 m) **Main rotor diameter:** 44' (13.41 m)
Cruising speed: 141 mph (227 km/h) **Useful load:** 3,377 lb. (1,531 kg) crew of two in tandem seats

Earlier single-engined models (upper sketch) show a **single large exhaust tilted up behind engine; all have flat-glass cockpit panes.** All have a distinct **pointed nose with prominent chin-turret.** Twin-engined Super Cobras, the Marines' vehicle of choice, have **large twin engine pods, cheek bulges for avionics, and rounded glass cockpit.**

An attack helicopter developed in the 1970s, with many upgrades, including dual engines, more armor, marine avionics, large TOW missile pods. Also prominent in Desert Storm operations.

The twenty-first-century Cobras show some important differences: **Horizontal stabilizer now has swept up-and-down end plates; tail rotor has shifted to the left side.** Side by side with earlier Super Cobras, the AH-1Z (lower sketch) has **perceptibly larger engines.**

McDonnell Douglas AH-64 Apache
Length: 58'3" (17.76 m) **Main rotor diameter:** 48' (14.63 m) **Cruising speed:** 184 mph (296 km/h)
Useful load: 3,685 lb. (1,671 kg) pilot and copilot/gunner in tandem seating (pilot in rear)

Two Army attack helicopters are seen near bases in the United States. This one has wheels (reversed fixed tricycle gear); the Bell AH-1 HueyCobra (preceding entry) doesn't. The Apache can fly upside down, a stunt sometimes performed at air shows. Otherwise, it has **huge armored engine nacelles over stubby outrigger wings, an angular greenhouse cockpit, bumps and lumps** for avionics and weapon systems, and rarest of all helicopter gear, a **tail wheel.**

In service since the mid-1980s, and a very successful machine in Operation Desert Storm, the Apache looks like hell on wheels when stripped; when armed with rocket launchers, like hell itself. Like the MBBs, it can fly in negative g; that is, it can push itself down.

New AH-64Ds (lower drawing) and refitted earlier AH-64s show a distinct fire-control radome over the main rotor.

Bell OH-58D Kiowa, SeaRanger

Bell AH-1 HueyCobra, SeaCobra

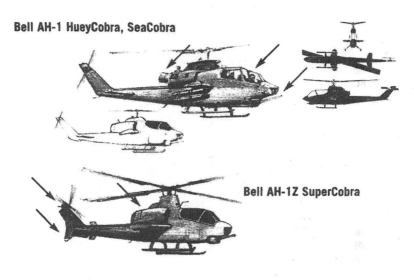

Bell AH-1Z SuperCobra

McDonnell Douglas AH-64 Apache

Boeing AH-64D Apache Longbow

Kaman H-2 Seasprite, SH-2G Super Seasprite
Length: 52'7" (16.03 m) **Main rotor diameter:** 44' (13.41 m) **Cruising speed:** 138 mph (222 km/h)
Useful load: 5,070 lb. (2,300 kg) two pilots and a sensor operator

Very rare; never seen in North America except in Navy paint. Odd cutback vertical stabilizer with very small horizontal stabilizers, twin engines with unusual side venting and a tapering nacelle.

An old warhorse, its basic mission is antisubmarine warfare. When operational, it fairly bristles with up to a dozen missiles and several torpedoes and even a depth charge. It is fully operable with only one engine and in 2004 was acquired by Australia, New Zealand, and Egypt. The Super Seasprite has long-life composition rotors and engine upgrades.

Sikorsky S-62, HH-52A
Length: 45'5" (13.86 m) **Main rotor diameter:** 53' (16.16 m) **Cruising speed:** 98 mph (158 km/h)
Useful load: 3,017 lb. (1,368 kg) crew of 2 and 10 passengers

Rare, declining numbers in Coast Guard red and white; scattered offshore oil use. Single air intake directly over cockpit; floats on outrigger braces; four symmetrical side windows; single horizontal stabilizer on port side; boat-type hull.

First delivered to the U.S. Coast Guard in 1963, it is being phased out in favor of more reliable twin-engine craft in both government and private service. Its amphibious ability (no extra flotation devices required) made it the choice, in the 1960s, for transporting passengers between San Francisco and Oakland airports.

Sikorsky S-61, SH-3 SeaKing
SeaKing data: Length: 72'8" (22.15 m) **Main rotor diameter:** 62' (18.9 m)
Cruising speed: 136 mph (219 km/h)
Useful load: 8,635 lb. (3,618 kg) military crews vary; civilian versions carried up to 30 passengers

Both commercial and military versions have a boat-shaped hull, but civilian versions were not all waterproofed. SeaKings have flotation pods on outriggers. Civilian versions showed up to nine side windows.

These were the giant helicopters, in civilian dress, that used to rattle Manhattan's windows between the Pan Am building and La Guardia Airport. They were the first Navy helicopters capable of searching for, and carrying the weapons to destroy, submarines.

Sikorsky S-61R HH-3 Pelican, Jolly Green Giant
Length: 73' (22.25 m) **Main rotor diameter:** 62' (18.9 m) **Cruising speed:** 144 mph (232 km/h)
Useful load: 8,795 lb. (3,990 kg) normally, crew of 2 plus engineer/flight chief; up to 30 troops

If the big red stripe doesn't convince you that this is the Coast Guard's Pelican, note the slim tail boom, larger flotation pontoons, and the sharp cutaway for hydraulic doors at the rear of the fuselage. In camouflage color, it's an Air National Guard or Air Force Reserve Jolly Green Giant.

Kaman H-2 Seasprite

Sikorsky S-62, HH-52A

Sikorsky S-61, SH-3 SeaKing

Sikorsky S-61R HH-3 Pelican, Jolly Green Giant

Sikorsky S-70, UH60 Blackhawk, CH-60 Seahawk, Jayhawk
UH60-A data: Length: 64'10" (19.76 m) **Main rotor diameter:** 53'8" (16.36 m)
Cruising speed: 167 mph (268 km/h)
Useful load: 10,716 lb. (4,861 kg) crew of 2, gunner, and up to 14 troops

Note fixed reversed tricycle landing gear; tail wheel midway down tail cone (closer to fuselage on Seahawks); airplane-looking tail unit, tail rotor canted to starboard. The side-window pattern is distinctive: two small rectangles forward, two large squares in sliding "barn door" entryway.

Thousands of Blackhawks at U.S. and overseas airborne infantry bases; hundreds more Seahawks at naval bases and aboard frigates, cruisers, and destroyers, where they provide air defense radar, antisubmarine capability, and rescue service. Designed in the late 1970s for carrying troops into combat, the Blackhawk and its derivatives now mount weapons, deliver mines, and provide both target acquisition and radar suppression for fixed-wing fighter-bombers. First used as the president's personal helicopter in 1989.

Sikorsky S-65, CH-53 Sea Stallion, Super Jolly Green Giant
Stallion data: Length: 88'3" (26.9 m) **Main rotor diameter:** 72'3" (22.02 m)
Cruising speed: 173 mph (278 km/h) **Useful load:** 19,556 lb. (8,870 kg) crew of 3 and 37 troops

Twin engines mounted away from fuselage; blunt nose; high cabin windows; large sponsons; odd one-sided horizontal stabilizer projects to starboard. Later models have canted vertical stabilizer, as does Super Stallion (next entry).

Many fewer CH-53s fly than the Hawk series. Most common near Marine Corps bases. Although built as a large and durable troop carrier, the CH-53 has a very high ceiling and is quite maneuverable. Several have been purchased by Alpine countries as mountain rescue craft. In Navy paint, serves as a transport, rescue platform, and minesweeper.

Sikorsky S-80, CH-53E Super Stallion, Sea Dragon
Length: 99' (30.18 m) **Main rotor diameter:** 76' (24.08.m) **Cruising speed:** 173 mph (278 km/h)
Useful load: 30,000 lb. (13,607 kg) crew of 3 and up to 55 troops

Huge. It's the largest helicopter outside Russia. Sponsons on Marine Corps Super Stallion are larger than on Sea Stallion; Navy's minesweeping Sea Dragon (illustrated) has **enormous fuel-holding sponsons.** Projecting to starboard, the **horizontal stabilizer is a gullwing. Three engines** (third one is on port side, above and behind the others); at rest, **seven-bladed main rotor; vertical tail fin canted to port.**

So totally unlike the CH-53 series Sea Stallions, the Super deserved a military model number of its own, not simply the suffix E. Has double the lift capacity, whether you're counting troops or tons, and four times the range of the earlier CH-53s. The Sea Dragon is also used by the Japanese military for submarine hunting and as a minesweeper. The canted tail fin (now being retrofitted on non-Super models) acts as an airfoil, increasing range and lift at altitude.

266

Sikorsky S-70, UH60 Blackhawk, Seahawk, Jayhawk

Sikorsky S-65, CH-53 Sea Stallion, Super Jolly Green Giant

Sikorsky S-80, CH-53E Super Stallion, Sea Dragon

267

Sikorsky S-92 Helibus
Length: 56'10" (17.32 m) **Main rotor diameter:** 58'1" (17.71 m) **Cruising speed:** 165 mph (265 km/h)
Useful load: 9,700 lb. (4,400 kg) crew of 2 and 19 passengers

This very large transporter is an amalgam of Sikorsky's earlier and even larger military helicopters: **a single bent and braced horizontal stabilizer** (see the Sikorsky S-65 and S-80, page 266) **on the port side of the tail fin** in this case; **side-mounted, exposed, turboshaft engines that begin in line with the center of the main rotor** (as on the Sikorsky S-70 Blackhawk, page 266); **large sponsons each side carry landing gear** (S-65 and S-80 style); and a **Blackhawk-shaped nose with much more glass.**

This all-new helicopter gives the external appearance of older Sikorsky designs but is essentially different, incorporating a good deal (40% by volume) of composites in the fuselage, as well as a new rotor blade design that provides more lift per horsepower. It has been extremely slow getting into production; design began in the early 1990s, and certification was awarded in 2004.

Kazan Industries (Russia) Mi-17, Mi-8
Length: 59'9" (18.22 m) **Main rotor diameter:** 69'10" (21.29 m)
Cruising speed: 140 mph (225 km/h) 2 pilots and up to 24 combat troops

Extremely rare but does fly in the Mexican self-defense force as an anti-insurgency gunship. **Long tail boom, low-mounted horizontal stabilizers, and a strongly sloping tail fin. Twin engines over the cockpit, with side exhaust near the rotor base. Five porthole windows each side. Always with fixed gear.**

This robust Soviet-era troop carrier is scattered worldwide, and "surplus" Mi-8s are ferrying salmon fishermen to remote rivers in Siberia and along the shores of the White Sea in eastern Russia. The unarmored versions, when normally loaded, can operate on only one engine.

Sikorsky S-64 Skycrane, CH-54 Tarhe
Length: 88'6" (26.97 m) **Main rotor diameter:** 72' (21.95 m) **Cruising speed:** 109 mph (175 km/h)
Useful load: 22,400 lb. (10,160 kg) crew of three, with triple controls

Nothing else looks like it. Sharply **cut-out fuselage** is for carrying encapsulated cargo; **long rear "legs" to landing gear.**

A packaged-load-lifter, the Skycrane program included a "universal military pod" for cargo or troops and a removable winch system capable of lifting 15,000 lb. (6,800 kg). A few used in civil works, including lifting powerline pylons and delivering bulldozers to isolated sites. Among other signs of lifting power, an unloaded CH-54 has reached an altitude of 11,000 ft. (3,353 m). Used extensively in the Vietnam War to move heavy equipment and armored vehicles and to retrieve crashed airplanes.

Sikorsky S-92 Helibus

Kazan Industries (Russia) Mi-17, Mi-8

Sikorsky S-64 Skycrane, CH-54 Tarhe

269

Boeing Vertol/Kawasaki 107, CH-46 Sea Knight

Length: 44'10" (13.66 m) **Main rotor diameter (each):** 51' (15.5 m)
Cruising speed: 155 mph (249 km/h) **Useful load:** 9,933 lb. (4,506 kg) crew of 2 and 25 troops

One of two double-ended helicopters still flying. Sea Knight is smaller than Chinook (next entry) and has large sponsons just forward of the tail, tricycle landing gear, and four round porthole side windows in rectangular frames.

The Sea Knight was the first replacement for the Flying Banana of the Korean War era, but the Army wanted a larger version and opted for the Chinook. After considerable upgrading and refitting, the Sea Knight was and remains the U.S. Marines' standard medium-size assault helicopter. A few in Navy paint as cargo carriers; even fewer in civilian dress in the United States. Under license, Kawasaki manufactures a military version for Japan and the only civilian models.

Boeing Model 234, CH-47 Chinook

Latest model CH-47D data: Length: 99' (30.18 m) **Main rotor diameter (each):** 60' (18.29 m)
Cruising speed: 154 mph (248 km/h) **Useful load:** 14,160 lb. (6,423 kg) crew of 3 and up to 55 troops

Compared to the Sea Knight (preceding entry): continuous bulge along lower sides of fuselage; engines exposed below rear rotor; five round portholes, fixed four-wheel landing gear. Much rarer civilian versions show more than a dozen rectangular passenger windows on each side.

The workhorse Army troop carrier of the Vietnam War, the Chinook was also used as a heavy-cargo transport. More than 400 older Chinooks are being converted to the 47D standard, which almost doubles the useful load and range of the Vietnam-era copters. Civilian versions are in service in the Far East, at both Gulf of Mexico and North Sea ports as airbuses for platform workers, and in the Pacific Northwest as airborne fire engines.

Boeing Vertol/Kawasaki 107, CH-46 Sea Knight

Boeing Model 234, CH-47 Chinook

Kaman K-MAX

Length: 41'9" (12.73 m) **Rotor diameter:** 48'4" (14.73 m) **Cruising speed:** 100 mph (160 km/h)
Useful load: 6,000 lb. (2,720 kg) pilot; for passengers, read following

A rare helicopter that combines **side-by-side twin pylons,** each with a rotor moving opposite the other (counterrotating), and tilted so that the blade clears the other pylon. Much smaller Kaman H-43 Huskies (next entry) have the same rotor system. The single-mast MD helicopters (pages 242–244) also lack tail rotors but otherwise look nothing like a Kaman.

A very rare but very noticeable sky crane. The dual and counterrotating rotors eliminate the need for the typical torque-balancing tail rotor found on other helicopters. If by the oddest chance you should see one with two passengers perched outside over the main landing gear, your eyes do not deceive. Ordinarily seen without passengers while engaged in logging, off-road construction, and, possibly, film clips from Iraq, where it has been used to deliver material from ship to shore.

Kaman H-43 Huskie

Length: 25'2" (7.67 m) **Main rotor diameter (each):** 47' (14.33 m)
Cruising speed: 98 mph (158 km/h) **Useful load:** 3,800 lb. (1,360 kg) crew of 2 and up to 10 troops

Rare. **Twin counterrotating main rotors, no tail rotor, complicated double-finned tail assembly.**

Unless you are near a Russian naval fleet on maneuvers, this is the only other twin-rotor helicopter you will ever see. The Russian naval helicopters have both of their counterrotating main rotors mounted on a single mast and have tails resembling the Huskie's. Huskies were once seen at every U.S. Air Force base, where they were used as firefighting and rescue craft. The Huskie was a common sight in Vietnam; now in civilian use only and extremely rare.

Groen Brothers Aviation Hawk 4 Gyroplane

Length: 24' (7.32 m) **Main rotor diameter:** 42' (12.8 m)
Useful load: 1,030 lb. (467.2 kg) pilot and three passengers

If you see a **twin-boomed "helicopter" with an added-on pusher-prop and no tail rotor** of any kind, you are looking at a Groen Hawk 4.

The world is replete with small home-built gyroplanes (about as accident-prone as other home-builts, such as motorized hang gliders), but there hasn't been a commercially successful large gyroplane built since World War II. The postwar attempts used helicopter blades and rotor heads to achieve FAA certification. A Groen Hawk was used successfully as a security observation platform at the Salt Lake City Winter Olympics without any trouble and no downtime for maintenance. Essentially, the freely rotating blades provide lift, and the pusher-prop provides forward motion. With the prop moving enough air, the Hawk can take off, in ideal conditions, nearly vertically and land vertically. In higher altitudes and loaded, it can still get off with less than 50 ft. (15.24 m) of roll.

Kaman K-Max

Kaman H-43 Huskie

Groen Brothers Aviation Hawk 4

273

Sikorsky S-55, H-19 Chickasaw
Length: 57' (17.39 m) **Main rotor diameter:** 48' (14.63 m) **Cruising speed:** 90 mph (145 km/h)
Useful load: 1,795 lb. (770 kg) pilot and up to 10 passengers

A very rare antique. **High cockpit; look-down windows well back on side of cockpit; four-wheel fixed landing gear standard; row of cool-air intakes below cockpit windows; deep fuselage with peculiar bracing fillet marrying fuselage to tail boom.**

This was the first commercial helicopter licensed in the United States. A few hundred civil and a few thousand military versions were built before 1955. It survived for years in the military of small South American countries and may still be seen in utility roles in rural America.

Sikorsky S-58T, CH-34 Choctaw
Length: 65'10" (20.06 m) **Main rotor diameter:** 56' (17.07 m) **Cruising speed:** 127 mph (204 km/h)
Useful load: 5,725 lb. (2,596 kg) pilot and 16 passengers

Quite rare; nonturbo models (lower sketch) are probably parked for the duration. **High cockpit; long, low, straight fuselage merging into tail cone.** Side windows vary; **six most common; twin "nostril" air scoop for turboshaft engine.**

Successor to the S-55, built with piston engines in the United States during the 1950s. The British-built Westland S-58s had turboshaft engines from the beginning, an idea taken up a decade later here. Although all S-58 production stopped in the early 1960s, Sikorsky continued turbo-conversion sales into the late 1970s. About 100 still flying, most now in utility rather than passenger roles. Common troop carrier early in the Vietnam War.

N

Sikorsky S-55, H-19 Chickasaw

Sikorsky S-58T, CH-34 Choctaw

275

Boeing Vertol and Bell Helicopter V-22 Osprey

Length: 57'4" (17.47 m) **Rotor span:** 83'10" (25.6 m) **Cruising speed:** 317 mph (509 km/h)

Still rare. When flying, most often seen in northern Virginia. The craft is bulky, bumpy, and oddly curved; twin outboard tail fins; two huge rotors (38 ft. [11.6 m] in diameter). An ungainly upsweep in the fuselage from the bulbous main landing gear fairing to the tail.

This aircraft, which first flew in 1989, is in both the developmental stage and the production stage, a first in U.S. military history. It has been notoriously unsafe: four crashes and 24 dead by 2004. It was rejected as "unsuited" by the Marines who provide helicopter service to the executive branch. The V-22 was supposed to be able to deliver itself transoceanically to any battlefield in the world. That appears an unlikely achievement. Derided by liberals and conservatives alike (the Cato Institute is a harsh critic). The Marines, however, are keeping the program alive (it was supposed to be adopted by all services) in spite of the problems. *Semper fidelis.*

Bell/Agusta Aerospace BA609 Tiltrotor

Length: 44' (13.31 m) **Wingspan:** 32'10" (10 m) **Cruising speed:** 290 mph (466 km/h)

Except for the military V-22 Osprey (still under development), this is the only aircraft with a pair of propellers/rotors at the wingtips. On the ground, the propellers will be horizontal to the ground; in flight, in normal propeller mode at right angles to the earth's surface.

The BA609 can be seen test-flying near Fort Worth, Texas, and it will be a miracle if it is seen elsewhere. It has the vertical takeoff and landing capability of a helicopter, combined with the reasonable speed of a small twin-turboshaft airplane. Although the military Osprey is riddled with problems, the BS609 may fly: It is much lighter, and because it is not designed to have folding wings for aircraft-carrier deployment, it has the virtue of relative simplicity. Bell/Agusta hopes to receive certification in 2007.

Boeing V-22 Osprey

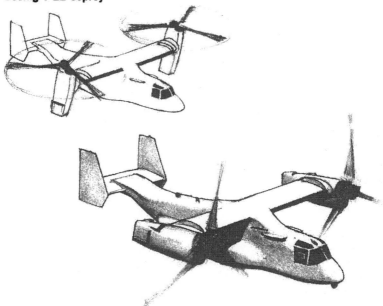

Bell/Agusta BA609 Tiltrotor

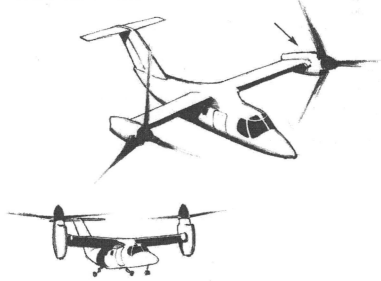

INDEX